Praise for

CLOSE TO HOME

"Thor Hanson has done it again! In this surprising, upbeat, and exciting book, we learn that not only everyday wonders, but actual scientific discovery awaits us close to home. The perfect mix of science and story, told by a master of both. I loved it!"

—Sy Montgomery, author of *The Soul of an Octopus*

"Packed with inspiration and insight, the wonders of the living world are vividly revealed in this beautifully crafted invitation to curiosity and exploration."

—David George Haskell, author of *The Forest Unseen*

"Thor Hanson's *Close to Home*, written with classic Hanson prose and enthusiasm, is as addictive as it is entertaining. Hanson teaches the value of looking high and low, near and far, at the fascinating natural systems we have never noticed right in our yards. And, very much to my liking, he tells us how to enhance those systems with minor tweaks to our landscapes. Thor Hanson has forced me to use the dreaded cliche: I couldn't put it down!"

—Doug Tallamy, author of *Nature's Best Hope*

"Thor Hanson's *Close to Home* invites us to step into the ordinary and discover the extraordinary. From natural science experiments to the search for new species, Hanson reminds us that awe and wonder are as close as our own backyard. With Hanson as our guide, a walk around the block becomes an opportunity to explore, to conserve, to ask questions, and to broaden our horizons." —Amy Stewart, author of *The Drunken Botanist*

"When we love a book we typically say, 'I couldn't put it down.' *Close to Home* is different. This book inspires such a rush of curiosity, I simply had to put it down over and over again so I could run out the door and investigate: peer beneath leaves for never-known insects, or turn my ear to the subtle language of neighborhood birds, or reach my hands into the soil to touch the under-earth beings beneath my everyday notice. What joy. Thor Hanson's love for his subject is infectious, and reminds us that access to the deep knowledge we need in this complex ecological time does not lie only within institutional walls, but in the everyday conversations between our wild minds, our wild hearts, and the wild earth."

—Lyanda Lynn Haupt, author of *Rooted*

CLOSE TO HOME

Also by Thor Hanson

Hurricane Lizards and Plastic Squid

Buzz

The Triumph of Seeds

Feathers

The Impenetrable Forest

For Children

Star and the Maestro

Bartholomew Quill

CLOSE TO HOME

THE WONDERS OF NATURE JUST OUTSIDE YOUR DOOR

THOR HANSON

BASIC BOOKS
New York

Basic Books
Hachette Book Group
1290 Avenue of the Americas, New York, NY 10104
www.basicbooks.com

Printed in the United States of America

First Edition: March 2025

Published by Basic Books, an imprint of Hachette Book Group, Inc. The Basic Books name and logo is a registered trademark of the Hachette Book Group.

Print book interior design by Amy Quinn.

Library of Congress Cataloging-in-Publication Data
Names: Hanson, Thor, author.
Title: Close to home : the wonders of nature just outside your door / Thor Hanson.
Description: First edition. | New York : Basic Books, 2025. |
Includes bibliographical references and index.
Identifiers: LCCN 2024016575 | ISBN 9781541601246 (hardcover) |
ISBN 9781541601260 (ebook)
Subjects: LCSH: Urban ecology (Biology)—Popular works.
Classification: LCC QH541.5.C6 H36 2025 | DDC 577.5/6—dc23/eng/20241217
LC record available at https://lccn.loc.gov/2024016575

ISBNs: 9781541601246 (hardcover), 9781541601260 (ebook)

LSC-C

Printing 1, 2024

For Dave Barrington and the late Steve Brunsfeld,
outstanding scientists, generous mentors,
and excellent friends

Contents

Map		*viii*
Author's Note		*xi*
INTRODUCTION	Backyard Biology	1

Part 1: Seeing

CHAPTER ONE	Look at Your Fish	13
CHAPTER TWO	Get Small	29

Part 2: Exploring

CHAPTER THREE	Something New	51
CHAPTER FOUR	Forces of Habit	71
CHAPTER FIVE	Above	89
CHAPTER SIX	Below	109
CHAPTER SEVEN	All Wet	129
CHAPTER EIGHT	After Hours	149

Part 3: Restoring

CHAPTER NINE	The Welcome Mat	169
CHAPTER TEN	The Limiting Factor	191
CONCLUSION	Wild Crescendo	211
Acknowledgments		*223*
Appendix: Citizen Science Resources		*225*
Notes		*227*
Bibliography		*253*
Index		*275*

A map of the author's yard. Illustration © Chris Shields.

Author's Note

I hope readers will take time to peruse the Notes section at the end of this book. It includes many backyard biology tidbits that didn't quite fit the narrative, but that nonetheless enhance the discussion, such as stargazing dung beetles, the bird chasers of Rome, and the reason there would be no pesto, mint tea, or medicinal marijuana if the world didn't abound with glandular trichomes.

Also, I would like to make two points about vocabulary. Many chapters in this book feature examples of "citizen science," collaborations between professionals and laypeople that can transform everyday observations into cutting-edge research. Though the movement is widely embraced, the term itself has come under criticism, leading to the proposal of more inclusive alternatives, such as "community science" or "participatory science." With no replacement yet agreed upon, I have elected to maintain the traditional phrase for the sake of clarity. Related issues of inclusion also apply to the term "backyard biology," with its implication that anyone and everyone has ready access to a yard. But the word "backyard" can also be applied more broadly, to any place near where people live—a neighborhood, a local landscape, or even a home region. Nor is the traditional concept of a "yard" necessary for meaningful nature exploration. I know someone who spotted an unusual leafcutter bee nesting on her porch, and I have another friend who found a rare flower mutation on a potted plant *inside* a tiny urban apartment. I use the phrase "backyard biology" with its most general definition in mind, celebrating stories that emphasize *curiosity*, an essential ingredient for connecting with nature anywhere.

INTRODUCTION

Backyard Biology

Who then persuaded you to stay at home?

—William Shakespeare
Henry IV, Part II (c. 1597)

It began with a bird and a window. The bird was a hermit thrush, and the window was the one that occupies the northern wall of my office shack. Their interaction ended in the predictable way: a thump, and then silence. Rushing outside, I found a small brown body lying in the grass, still warm to the touch. I'd never seen a hermit thrush up close before. Its muted brown plumes looked perfectly suited to a life spent skulking in thickets—hard to spot, except for the tail, which glowed with tones of russet, like newly varnished wood. Draped across my palm, those few feathered ounces felt utterly insubstantial, more like the idea of a bird than something so recently alive and flitting.

As I laid the thrush to rest beneath a rosebush, my sorrow triggered something more than regret. I also felt chagrined. Here I was, studying nature and writing books about it, and I'd had no idea that this celebrated bird was wintering in the shrubs just a few feet from my desk. Biologically, the hermit thrush is famous for its song, a cascade of two-note phrases arranged in precise harmonic intervals, eerily similar to the minor chords and scales used in human music. Male birds often sing at dusk on spring and summer evenings, filling the air with strains so haunting they've earned the species a prominent place in literature. By one count, well over 150 poets and essayists invoked the hermit thrush in their work between 1860 and 1940, including such well-known figures as T. S. Eliot and Walt Whitman.[1] For Whitman, only "the wondrous chant of the gray-brown bird" could both convey and soothe the national grief at the center of "When Lilacs Last in the Dooryard Bloom'd," his elegy to Abraham Lincoln.[2] As a biologist and an author, failing to notice a species so well-known in both of my chosen trades begged an obvious question: What else was I overlooking in my own backyard? Finding the answer inspired this book, and also allowed me to take up an old challenge.

My wife's grandparents lived for decades in a shoreline cabin built from beach logs, where the family still gathers for holidays and special occasions. Among the keepsakes and photos pinned to the wall, a faded, handwritten note reads: "Environmentally, the most radical thing you can do is stay home." It was written long before anyone worried about the impact of travel on their

FIGURE I.1. The hermit thrush is famed in biology and literature for its song, an ethereal cascade of notes that nature writer John Burroughs called "the finest sound in nature." John James Audubon, *The Birds of America* (1827). Courtesy National Gallery of Art, Washington, DC.

carbon footprint, hinting instead at deeper lessons from an earlier, more rooted generation. Home life once involved a lot more time spent outside, even in urban areas—from children's play to yard work and gardening, walks in the park, or simply hanging laundry in a courtyard. More than one psychology degree has since been earned studying the effects of such daily outdoor activities, and how they deepen people's connections to nature.[3] Whoever wrote that old note was onto something, which may help explain the results of a recent experiment, one that took place simultaneously, and unintentionally, in ecosystems all over the world.

During the early months of 2020, lockdowns triggered by the COVID-19 pandemic forced pretty much everyone to stay home, sometimes for months on end. And the environment responded. Smoggy skies cleared over major cities; fish returned to the canals of Venice; roadsides bloomed in the absence of mowing; and wildlife sightings ticked up everywhere, from whales in the harbors of Vancouver and New York to wild boars roaming the streets of Barcelona. News headlines celebrated these signs of nature's resilience, and studies soon linked them to the sudden drop in traffic, commerce, industry, and other modern pursuits. Fewer beachgoers in Florida and Brazil, for example, allowed sensitive shorebirds and sea turtles to expand their nesting habitat, while fewer cars on highways reduced road deaths for everything from mountain lions in California to hedgehogs in Poland to wallabies and brushtail possums in Tasmania. White-crowned sparrows in San Francisco sang more quietly as city noises ebbed, and humpback whales altered their vocal patterns in Alaska's Glacier Bay, where they no longer had to compete with the underwater clamor of cruise ships and yachts. It seemed that wherever human activity diminished, nature adjusted and filled in the gaps, quickly taking advantage of a caesura that biologists began calling "the anthropause."[4] But staying home did more than facilitate this moment of global rewilding. It also put people in a position to notice.

From patios and gardens to balconies, rooftops, and neighborhood parks, local outdoor spaces became essential sources of respite during the pandemic. And the more time people spent outside, the more they began to see. Yes, some species behaved differently during the anthropause, but many of the most notable wildlife reports had more to do with awareness—people tuning in to things that had been there all along. Photos of surprising backyard flora and fauna flooded social media, so many that scientists began mining the images for usable data! But those discoveries were more than just a pandemic fad; they tapped into a knack for nature observation that people once took for granted. It wasn't a novelty or a learned skill; it was part of being human.

Anthropologists curious about how the environment shaped human evolution often look for analogies in the lives of modern hunter-gatherers. Though culturally and technologically distinct from our ancestors, any people subsisting off the land with simple tools reflect that shared, ancient lifestyle. They face the same basic challenge of extracting sustenance from their immediate surroundings, a task that demands intimate and hyperlocal biological knowledge.[5] Typical San hunters in the border region of Botswana and Namibia, for example, can identify over 220 different animals and 150 kinds of edible and medicinal plants, all within a half-day's walk of camp. In the eastern Congo Basin, botanical knowledge amassed by the Mbuti and Effe peoples includes the names and uses of nearly 800 rainforest species, ranging from the obviously handy (edible

FIGURE I.2. The author in Uganda's Bwindi Impenetrable National Park in 1994, hands festooned with butterflies and tiny stingless bees.

fruits and berries) to the downright arcane (leaves known to occasionally attract tasty caterpillars). But the human potential for fluency in nature is hard to fully grasp from facts and figures. My own appreciation comes in no small part from an experience in Uganda, where a colleague's research project was saved by a local man who casually accomplished something the rest of us couldn't fathom: tracking bees in a jungle.

In the mid-1990s, I worked on a mountain gorilla project in Uganda's Bwindi Impenetrable National Park. The job involved a lot of hiking in hot weather, and we were often set upon by hordes of insects attracted to our sweat, including tiny, jet-black bees no larger than a pencil point. The forest abounded with at least a dozen species of these gentle, stingless creatures. Yet when an entomology graduate student arrived to study them, he spent days fruitlessly searching for their nests (and nights frantically worrying about the fate of his master's thesis). The park rangers and I suggested that he consult the Batwa, a group of hunter-gatherers who, until quite recently, had called that forest home. He took our advice and described to me later how a Batwa man had led him to a small clearing in the forest where they settled down beside some flowers and waited. And waited. And just when he thought, "What am I paying this guy for?" a bee showed up, messed about on a flower, and zipped off again. They waited some more, and watched as that process repeated itself several times. Then the man stood up and walked briskly into the forest for a short distance, before stopping, looking around, and pointing to a hole in an old stump, where the bees were busily going in and out.

How did he do it? Well, he knew what flowers the bees preferred, so he knew where to wait. He watched their flight path, so he knew the direction of the nest, and he saw them come and go several times, so he knew how long it took to make a round trip. Since he also had a good idea of how fast bees fly, he was able to calculate the distance. Put those things together, and it was a simple matter of pacing off the right number of steps in the right direction, and then looking for the sort of rotten log or stump where he knew the species in question liked to nest. For a trained entomologist, this had been a baffling, unsolvable challenge. For a lifelong forest dweller, it was child's play.

Decades later, and in spite of a career in field biology, that story still astounds me. I think of it as a benchmark of the possible, what people are capable of when observation is immersive and habitual, a state of being rather than an occasional effort. It is also a reminder that our drift from those skills is largely circumstantial, the result of living mostly indoors, focusing our attention on media and technology. Wherever subsistence habits persist, people's senses in nature remain acute, and some level of that fluency is attainable for all of us. As a species, our collective history includes a lot more bee-tracking than texting. And that makes reacquainting ourselves with backyard biology more than just an exploration; it is also an act of remembrance.

Historically, scientists have not been blind to the benefits of studying nature close to home. Luminaries from Isaac Newton to Charles Darwin and Gregor Mendel all made important breakthroughs while working in their yards and gardens. But the vast majority of backyard observations have either gone unrecorded or been tucked away in personal journals and notebooks. Henry David Thoreau's invaluable records on the plants and birds at Walden Pond, for example, were kept apart from his books and essays, and remained unpublished and unknown to science for nearly 150 years.[6] In the twenty-first century, all that is changing. The rise of citizen science has freed backyard biology from the constraints of isolation and geography, allowing information from any patch of habitat to immediately contribute to large, cutting-edge research projects. Examples in these pages range from the discovery of new flies and fungi to the behavior of urban parrots, and online portals like SciStarter catalog thousands more. It's often said that knowing more about nature leads to caring more about it; or, put another way, action is curiosity's consequence. That means the plants and animals around us stand to gain as more and more scientists and backyard explorers reach the same conclusion: human-dominated landscapes can—and must—become better habitat if we are to counter the loss of biodiversity in our rapidly changing world.

This book follows the three themes that have been my guideposts, and that lie at the heart of backyard biology everywhere: **seeing** (how to look), **exploring** (where to look), and **restoring** (how to help). I offer it as both a report and a road map—not just to describe my own explorations, but to encourage yours. Paradoxically, that means that I hope that reading this book will inspire you to stop reading it, throw open the door, and head for the nearest green space you can find. Perhaps you will take the book with you, and choose to turn its pages in the same kind of setting where many of them were written: out in the open air, back propped against a familiar rock or tree trunk, all senses on high alert.

PART ONE

Seeing

The universe is full of magical things patiently waiting for our wits to grow sharper.

—Eden Phillpotts
A Shadow Passes (1918)

CHAPTER ONE

Look at Your Fish

The lesson which life repeats and constantly enforces is "look under foot." . . . Every place is under the stars, every place is the centre of the world.

—John Burroughs
Leaf and Tendril (1908)

A few days after signing the contract to write this book, I decided to take the long route to work. My office shed sits in our orchard, a few hundred feet from the house, but I get more time outside if I go up to the county road and circle back on a trail through the woods and old pastures that make up our rural property.[1] It was a sunny spring morning, and I hadn't gotten far up the driveway before my ears registered the voice of a male Hutton's vireo, repeating his single, burry whistle-note again and again with cheerful insistence. Usually a species found high in the forest canopy, this individual sounded much lower down, where mated pairs sometimes descend to nest. I made a mental note to come back at some point and investigate, and then continued on my way to work. Two steps later, it hit me: if my job was to study and write about things living in my yard, then *I was already working*.

That thought stopped me dead in my tracks at the very moment the vireo appeared, hopping through the branches of a willow tree directly above my right shoulder. Drab at first glance, it was the sort of bird whose beauty rewarded closer scrutiny—body feathers that shifted from gray to green in different light; black wings that flashed with ivory bars in flight; and thin, white rings encircling each eye with a look of perpetual surprise. I kept still and watched as it fluttered and darted from twig to twig, still singing, and then dropped into the leaning crown of a tall ocean spray bush, where it promptly vanished and went silent. Stepping quietly, I closed in, eyes fixed on the last place that I'd seen the bird. Beneath a cover of leaves, the shrub's fine branch tips were festooned with lichens—clasping pale *Parmelia*, pearly tufts of *Ramalina*, and strands of a greenish *Usnea* that hung like rustic tinsel. There, just out of reach overhead, my eyes finally made out what amounted to the first biology lesson of the morning: if you want to hide a nest in the middle of lichens dangling from a twig, then *build* the nest out of lichens dangling from a twig. It was a master class in camouflage, a deep cup woven from the very same materials that surrounded it. I could just make out the bird's bill topping the rim on one side, and the tip of its tail emerging from the other.

FIGURE 1.1. A Hutton's vireo in the open (left), and hidden on its cryptic nest (right), with only its bill visible. Images © Michael Noonan.

The rest of the body lay tucked out of sight within, concealing and protecting what was probably a full clutch of eggs. Backing away, I found a grassy place to sit, and settled in to watch.

Nothing happened. Minutes ticked by as the nest and its cargo of feather and shell bobbed and swayed in the breeze with no sign of further activity. Which was, of course, precisely what the vireo had in mind. To be successful, incubation should be dull, because the one activity likely to take place is exactly what the bird is trying to avoid: an attack. Sitting songbirds are inherently vulnerable, and their eggs are in even greater danger, targeted by everything from owls to opossums, snakes, ravens, weasels, and more. The allure of such an easy, highly nutritious meal can tempt almost anyone, even otherwise vegetarian creatures like deer and squirrels. As if to underscore that ever-present threat, I slowly realized that a noise I'd been hearing in the background belonged to perhaps the most notorious nest raider of all.

There it was again, a querulous, chirping purr, like a housecat with a sparrow stuck in its throat. But I knew that sound, and it wasn't feline or avian. It belonged to a baby raccoon, which meant that the vireo and its mate had chosen a particularly precarious place to nest. Rising quietly to my feet, I edged toward a nearby snag, the remains of a large fir tree that had snapped off in a windstorm. Its leaning trunk ended twelve feet up in a tangle of shattered wood that folded over and sheltered a dark hollow. There, I saw a pair of white-rimmed ears, a telltale mask, and two shiny black eyes peering out. The mother raccoon seemed unconcerned by my presence, or perhaps she was just tired. If hers was a typical litter, then there were three or four lively kits tussling about in the den beneath her, their mewling a reminder of their constant hunger. Nursing that brood required a lot of calories, and if she ever noticed the nest so conveniently placed on her doorstep, then the vireos' eggs would be gone in a heartbeat.

"I know something you don't know," I whispered, as we stared at one another. Then I turned and walked back to the driveway, choosing a different route so as not to direct her attention to the vireos and their vulnerable clutch. But in the weeks ahead I would become a regular visitor, watching from a distance as her rambunctious youngsters grew bold enough to

FIGURE 1.2. A mother raccoon peers from her den in the hollow of a fir snag in the author's yard. Image © Michael Noonan.

climb up beside her and peer out, their faces like tiny mirrors of her own. Once, I arrived at dusk just as she was departing the noisy den on a foraging trip, picking her way down the trunk alone, and ambling off with an air of what is hard not to describe as relief.

Fortunately for the vireos, the mother raccoon never caught on to their presence. As the days passed, I saw how the avian parents shared the responsibilities (and tedium) of incubation, discreetly changing places on the nest after shifts that sometimes lasted well over an hour. Their trade-offs followed a predictable routine—soft chirps from the approaching parent signaled the other to silently depart, leaving the nest briefly empty. Then the newcomer would alight on the rim and peer inside, as if counting the eggs, before finally settling in to begin its shift. For the male, maintaining an aura of constant stealth put him at odds with another vital impulse—the urge to vocally stake out his territory. More than once I saw him break into full-throated song while sitting on the eggs, rather defeating the purpose of all that careful camouflage. (This habit may help explain why the female seemed reluctant to leave him there for long—in my observations, his shifts averaged twenty-five minutes shorter than hers.) But even with such lapses, the vireos managed to successfully hatch out and raise up two healthy chicks. By chance I spotted them soon after their first flight—a pair of awkward fledglings hopping and flapping through the understory below the nest, natal down still clinging to the backs of their heads like tufts of unkempt hair. The parents were there too, darting in occasionally to provide encouragement in the form of a tasty grub or caterpillar. With the family now mobile, the birds had no further use for the carefully crafted nest and they abandoned it immediately, moving off through the undergrowth, past the raccoon snag, and out of sight.

Feathery and fleeting, my last glimpse of the vireos was just the sort of edge-of-vision activity that plays out constantly in shrubs and trees and tall grass all around us, so easily ignored in our habitual hurry through nature. It takes a conscious effort to stop and investigate those furtive stories, let alone the more subtle cues given off by plants or insects, or creatures that hide from us entirely. Biologists conduct field research in part to simply slow down, focus their attention, and dedicate specific time and energy to the task of careful observation. It's true that insights sometimes come in exotic locations, watching novel species acting out unexplained behaviors. But familiarity invites its own revelations. Some things have to be seen again and again before they reveal their secrets, a lesson driven home by a famous journey of biological discovery that didn't even happen outdoors. It took place in an empty laboratory, with a dead fish in a jar.

If someone were to publish a who's who of late nineteenth-century grasshopper experts, entomologist Samuel H. Scudder would stand at the top of the list. He also studied crickets and butterflies, and his books on the fossil insects of North America are still cited today. But in spite of his academic accomplishments, Scudder will always be best known for a short essay he wrote in tribute to his late mentor at Harvard University, the famed naturalist Louis Agassiz. It recounts their first meeting, when Scudder introduced himself as a recently enrolled student eager to pursue his passion for insects. Agassiz responded with enthusiasm, but then surprised the young Scudder by handing him a huge specimen jar that contained a fish, preserved in yellow alcohol. "Take this fish," Agassiz instructed, "and look at it; we call it a haemulon; by and by, I will ask you what you have seen."[2]

With that, Scudder was left alone in an empty laboratory, equipped with nothing more than the dead fish, a pencil, and a notebook. "In ten minutes I had seen all that could be seen in that fish," he recalled, and then time really began to drag. "Half an hour passed—an hour—another hour; the fish began to look loathsome. I turned it over and around; looked it in the face—ghastly; from behind, beneath, above, sideways, at a three quarters' view—just as ghastly."[3] Nonetheless, he jotted down a list of its fishy features and dutifully made his report. Agassiz was unimpressed.

"You have not looked very carefully," the professor admonished. "Look again! Look again!"[4]

Embarrassed, and determined to prove himself, Scudder went back to his task with new zeal, and kept at it for three long days, prodded periodically by Agassiz's repeated directive to "Look at your fish . . . look, look, look."[5] Slowly at first, and then with increasing speed, he started discovering "one new thing after another."[6] He noticed the symmetry of the fish's paired fins and organs, and found fringes on the arches of its gills. He sketched and studied it from every angle, discovering patterns and details in the scales, teeth, eyes, lips, pores in the head, and more. On the fourth day, Agassiz introduced another, closely related specimen, and soon Scudder was comparing, contrasting, and recognizing subtle differences among dozens of species. Before ever examining a single insect, he would go on to spend eight months mastering the anatomy and taxonomy of the entire haemulon family of fishes.

"This was the best entomological lesson I ever had," Scudder wrote, describing how it taught him how to truly see and assimilate *whatever* was right in front of him.[7] That skill, he explained, became the cornerstone of his long research career.

Metaphorically, it's not much of a leap from the lab of Louis Agassiz to the average backyard. Replace Scudder's fish with familiar objects in familiar surroundings, and the potential for insight from sustained inspection is much the same. But it's one thing to draw a general conclusion about the power of observation, and quite another to make specific biological discoveries. Just how much can a curious person really expect to learn by examining the nature close to home? Few examples provide a more inspiring answer to that question than the life of perhaps the most famous biologist of all.

Though famed for his voyage around the world on HMS *Beagle*, with its auspicious stop in the Galápagos Islands, Charles Darwin was actually a notorious homebody. From the day he stepped back ashore at Falmouth in 1836 at the age of twenty-seven until his death nearly half a century later, Darwin never left Britain again. Nearly all that time was spent at Down House, the comfortable country estate where he and his wife raised a large family, and where he pursued his studies through avid reading, extensive correspondence, and a seemingly endless variety of backyard experiments.[8] He tracked bumblebees. He hand-pollinated foxgloves. He transplanted ants from one nest to another. He dunked ducks in duckweed to see if it would stick to their feathers. (It did.) Darwin also bred pigeons, pears, and primroses, and he made midnight excursions to observe the behavior of earthworms. He germinated seeds extracted from pond muck, fish bellies, bird droppings, and the pellets coughed up by owls. He often searched the yard for bird nests and beetles, and he once tried feeding his own toenail clippings to a carnivorous sundew plant. (The sundew wasn't interested.) As one of Darwin's neighbors aptly put it, "Home was his experiment station, his laboratory, his workshop."[9]

Many of the backyard investigations at Down House got their start during Darwin's daily rambles along the "Sand-walk," a quarter-mile (0.4-kilometer) trail that led from the back of the kitchen garden along a shady fence line that bordered his neighbor's hayfield. Biographers have referred to it as Darwin's "thinking path," but he used the Sand-walk to *watch* nature just as much as he did to ponder it. Again and again, Darwin's trailside observations fed directly into his work on larger themes. When a weedy mustard sprouted from disturbed ground beside the path after a nine-year absence, he followed up with a series of trials and published "Vitality of Seeds," an early paper on the concept of soil seed banks. Hand-pollinating a trailside patch of orchids inspired two years of research into the coevolution between insects and flowers, culminating in the book *Fertilization of Orchids*.[10] The Sand-walk and surrounding meadows also provided ample testing grounds for Darwin's long fascination with earthworms—their habits, their intelligence, and the vital role they play in moving soil up from the depths. He continued his walks and observations even as his health declined, and the backyard worm studies would be his last, published in book form just six months before his death in 1882.

It's safe to say that Charles Darwin paid close attention to the plants and animals living in his yard. More than a pastime, that habit of constant scrutiny anchored his approach to science

FIGURE 1.3. Though famous for his travels in the Galápagos Islands, Charles Darwin spent most of his long career at home, conducting experiments in the yard and taking daily rambles on a path he dubbed the Sand-walk. *The Illustrated London News* (1887). The Print Collector/Alamy.

and provided boundless opportunities for testing and refining his ideas. It's probably also safe to say that if backyard observations helped Charles Darwin to better understand evolution, then they probably have a lot to teach all of us. Of course, not everyone enjoys the benefits of living on a Victorian estate, with acres of grounds and gardens maintained by a full-time domestic staff. But as humble or grand as any individual version of a Sand-walk may be, all hold the promise of discovery. And as I continued along my own path, leaving the vireos behind and heading up the driveway to the road, it occurred to me that one feature of my yard might have filled old Darwin with envy: the trees.

Tall Douglas firs blanket much of the island where I live, part of the great temperate rainforest and related woodlands that hug the coastline of the Pacific Northwest from the northern tip of California to the Alaska panhandle.[11] Darwin never glimpsed these cool, drippy landscapes—his route on the *Beagle* remained far to the south. But he ranked their austral counterpart, the "primeval forests" of Tierra del Fuego, as one of the most sublime sights he ever beheld.[12] And he was certainly aware that woods of a similar scale thrived in northwestern North America. The Royal Horticultural Society, co-founded by one of Darwin's uncles, had sponsored Scottish botanist David Douglas on a hugely successful plant-collecting trip to the region in 1824. Douglas returned after three years with the seeds of more than 200 species, including those of his namesake fir and several other conifers capable of growing 10 feet (3 meters) or more in diameter and reaching heights of over 300 feet (91 meters). "A forest of these trees," he wrote, "is a spectacle too much for one man to see."[13] Other explorers agreed. In one account from 1862, a strapping Royal Marine stationed near modern-day Vancouver, BC, wagered that he could fell a full-grown Douglas fir with an axe, and gave himself a week to finish the job. "But at the end of three days he found his hands blistered painfully, and the tree upright and almost uninjured as before."[14] The week expired with the fir still standing.

Trees of such magnitude were unheard of in rural England, and while Darwin surely enjoyed the copse of hazels, privets, and hornbeams he planted alongside the Sand-walk, there must have been times when he yearned for something larger and wilder. That thought helped me see our woods with fresh eyes—not just as a familiar backdrop, but as a rampant green force that towered over the driveway and the two-lane country road we live on, shrinking the sky to pale strips and sun shafts glimpsed through a geometry of branches, twigs, and needles. I walked in that fractal shade to a narrow footpath that descended from the road across the far side of our land, weaving through tree trunks and shoulder-high thickets of salal, an evergreen shrub related to heather and rhododendron. Moss softened every footstep and draped the furrowed ridges of the fir boles where their bases flared to meet the ground. The hushed forest felt timeless, but for trees known to live a thousand years, these were still adolescents. The oldest among them dated to the late nineteenth century, part of the first generation to sprout after settlers and loggers cleared nearly all the region's original forest, rather overzealously finishing the job begun by that axe-wielding marine. Still, the firs in our backyard had put their time to good use. Several already topped out at more than ten stories tall, with trunks 3 feet (0.9 meters) across—a vertical mass of life and habitat that I realized, as later forays in this book will show, I knew virtually nothing about.

At a brisk pace, I can get from the road through our woods and out into the field below in less than four minutes. But I forced myself to slow down and concentrate on the details—the darker bark of the occasional lodgepole pine amongst the firs, or the spreading boughs of a young hemlock rooted in the remnants of a rotten stump. Splashes of red and pink announced the flowers of salmonberry and wild currant, and rough sword ferns brushed my legs, their long fronds arching from rosettes that crowded either side of the path. To a forestry professor I knew in graduate school, this scene would have qualified as "blah woods," the common, expected species growing together in common and expected ways. That phrase always struck me as jaded, but in the context of backyard biology it's downright dangerous. It not only devalues the beauty of everyday plants and animals, it blinds us to the many mysteries they still contain.

At the base of the slope, I crossed an old drainage ditch choked with slough sedge—a band of leafy green tufts that marked the water's path like a thirsty hedgerow. Beyond the ditch, coniferous trees gave way suddenly to a brighter forest crowded with alders—dozens of pale, silvery trunks under a leafy canopy lush with springtime growth. There was a thicket of salmonberries, a few more trees, and then the trail emerged into an open pasture where Charles Darwin—or anyone else familiar with English meadows—might have felt right at home. European settlers had brought their forage grasses with them, and our field contained several varieties known to occur at Down House, including tall fescue, wild rye, and sweet vernal grass, the first species that Darwin ever identified. "I have just made out my first Grass, hurrah! hurrah!" he exulted at the time.[15] With the help of Catherine Thorley, his children's governess and a skilled naturalist in her own right, Darwin went on to catalog 142 different grasses and wildflowers in Great Pucklands meadow, a 13-acre hayfield just over the fence from the Sand-walk. The number of species in our pasture was anyone's guess—yet another glaring unknown along the path of my daily commute.

Skirting the edge of the field, I passed along the fence line of our garden and then across an overgrown, gravel-covered lot ripe for restoration. From there it was just a few more steps to the orchard, home to ten apple trees, three pears, two plums, a quince, and a peach, as well as eleven chickens and two noisy ducks. My office was there, tucked among the fruit trees, a simple, one-room structure we called the Raccoon Shack in honor of its former occupants. (With only one hollow tree on the property, anything with a roof or a crawlspace was valuable denning habitat.) My journey from house to office had brought me through places so familiar I thought I knew them intimately. After all, I had been living on this land for more than twenty years. But even the slightest effort to "look at my fish" would quickly put that notion to rest. Just like finding vireos and raccoons living right alongside the driveway, new discoveries began piling up whenever I reminded myself not to just take the long route, but to take it slowly. I found a wild bitter cherry tree in full bloom, nestled among the firs, and I noticed a nuthatch chiseling out a nest hole in the top of a dead alder. I spotted species of ferns and orchids that I had overlooked right alongside the path, and I learned that if you listen long enough, a raven will sometimes inadvertently croak the cadence to "Shave and a Haircut . . . Two Bits." All these observations expanded my knowledge of the yard, but I realized that they all involved plants and animals

I was already well acquainted with. In a sense, I was only seeing things that I already knew how to see. An adage paraphrased from French novelist Marcel Proust reads, "The real voyage of discovery consists not in seeking new landscapes, but in seeing with new eyes."[16] That's excellent advice for any backyard biologist, and there are many ways to go about it. Perhaps the most straightforward, and a good place to start the next chapter, is to simply use someone else's eyes.

CHAPTER TWO

Get Small

There are no little things. "Little things," so called, are the hinges of the universe.

—Fanny Fern
Ginger-Snaps (1870)

"What made that, Papa?" Noah asked, pointing down at a patch of sparse grass and clover by the side of the path.

At first I couldn't see what he was talking about, but when I squatted and looked at the bare soil between the plants, there it was: a tiny turret of dried mud. Thinner than a pencil and barely half an inch (one centimeter) tall, it looked like a rustic ceramic vase, curved near the top and with a slightly flared lip. Up close, I could see that it was built from tiny balls of dirt, stacked and somehow smoothed together into an artful whole. But that didn't help me in the slightest.

"I haven't got a clue, Noah," I said, and I meant it. It wasn't just the species that had me stumped; I couldn't tell him the family, the order, or even the phylum! A worm? An insect? A spider? Ignorance on that scale might be cause for embarrassment in some professions, but I felt only a growing sense of excitement. In biology, the unknown is every bit as important as the known. Without mysteries, there can be no discoveries.

"Let's put a camera on it!" I urged, and a quick dash to the Raccoon Shack supplied the necessary equipment. Once we had the little video recorder whirring away, propped up by pebbles a few inches from the puzzling turret, we continued on with the project that had brought us outdoors in the first place: an after-school game of badminton.

I often tell people that if you want to see more when you go out in nature, don't take field guides, take a child. Borrow one if you need to. They will find the salamanders and spider eggs and countless other things you've been missing for the simple reason that they haven't learned not to. Children see more in nature because they have yet to develop the full suite of filters that all of us rely upon to block out information. Just imagine trying to concentrate on this paragraph if you had no ability to disregard the other sights, sounds, and smells in your immediate vicinity. It would be paralyzing—a serious sensory disorder. Developing what psychologists call "selective attention" is an unconscious but vital skill in a world filled with media, technology, and other clamorous distractors vying for notice. Blocking those inputs is necessary, but it comes at a cost, because our filters will silence the song of a warbler just as readily as someone's annoying ringtone. In fact, natural signals often drop out first, since most of them weren't intended to communicate with us in the first place.

Awareness in nature takes effort, but it can be learned. Or perhaps a better word is *relearned.* There is ample evidence that natural settings stimulate the human brain differently than built environments—nature therapy is often prescribed to reduce stress and anxiety. But some experts also believe it can reinvigorate overtaxed powers of attention and observation, tapping into senses that our wild ancestors spent thousands of generations honing and relying upon.[1] With fewer barriers to overcome, children offer us a glimpse of that capacity. They have a keen ability to notice details—things that adults unconsciously ignore as extraneous. While that can make it hard for kids to stay focused on a single task (as any primary school teacher will attest), in nature it amounts to its own form of attentiveness. Biologists have known about this for centuries. Alfred Russel Wallace, the codiscoverer of evolution by natural selection, employed village children wherever he stopped during his ten-year exploration of Indonesia and Malaysia. In exchange for a few coppers, his "little corps of collectors" would bring in "bamboos full of creeping things," many of them new to science.[2] This allowed Wallace to gather specimens constantly, even on days when he was too weak with fever to venture far from his hammock. Similar tactics were apparently employed at Down House, where Charles Darwin's children habitually scoured the grounds and nearby countryside for interesting finds. "I have a vivid recollection of the pleasure of turning out my bottle of dead beetles for my father to name," Francis Darwin later wrote, "and the excitement, in which he fully shared, when any of them proved to be uncommon ones."[3] The excited father went on to submit those rare beetle observations for publication in *The Entomologist's Weekly Intelligencer*, graciously crediting the finds to "three very young collectors."[4] For French entomologist Jean-Henri Fabre, the debt owed to children ran even deeper—it was their knowledge that introduced him to his life's passion: the subtle wonders of a landscape called the *harmas*.

Fabre began his career as a small-town schoolteacher in Provence in 1843, with no formal scientific training and no clear path to pursue his enthusiasm for natural history. That changed one spring day when he decided to take his students outdoors for some hands-on lessons in geometry. They walked to the harmas, a gravelly plain outside of town studded with low shrubs and clumps of grass. There, he set the class to work with surveying equipment, laying out great triangles and trapezoids on the open, level ground. Almost immediately, biology intervened.

> If I sent one of the boys to plant a stake, I would see him stop frequently on his way, bend down, stand up again, look about and stoop once more, neglecting his straight line and his signals. Another, who was told to pick up the arrows, would forget the iron pin and take up a pebble instead; and a third, deaf to measurements of angles, would crumble a clod of earth between his fingers. . . . The polygon came to a full stop, the diagonals suffered. What could the mystery be?[5]

When confronted, the errant pupils revealed all. As Fabre put it, they "had long known what the master had not yet heard of, namely, that there was a big black Bee who made clay nests on the pebbles in the harmas."[6] Each nest contained a dab of strongly flavored honey, easily scooped out with a piece of straw. "I acquired a taste for it myself," he confessed, "and joined the nest-hunters, putting off the polygon till later."[7] From that beginning, Fabre would find ways to spend more and more time exploring the harmas and describing the lives of its smallest inhabitants, from the black mud bee that started it all to various

FIGURE 2.1. Like an expert bricklayer, the black mud bee (*Megachile parietina*) builds its nest from tiny pebbles cemented together with mud. It fills each small cell with a gluey mixture of pollen and nectar intended as food for its offspring, but easily robbed by young naturalists in search of a sweet treat. Illustration © Chris Shields.

spiders, flies, glowworms, beetles, scorpions, grasshoppers, and more.[8] His scientific output increased dramatically late in life, after he retired from teaching and devoted himself fully to a fenced-off patch of harmas that served, appropriately enough, as the backyard of his rural home.[9]

Had Fabre visited our yard, Noah's turret would no doubt have fascinated him, and he would have been just as eager as we were to identify its builder. He might also have envied our digital camera. Not unlike deploying a troop of sharp-eyed children, it allowed us to continue making observations while we attended to other things—badminton, chores, homework, dinner. Later that evening, we downloaded our footage and scanned through it for signs of activity. Minutes passed at high speed, the grasses and clovers twitching in invisible breezes. We saw two thatch ants scurry by, indifferent to the turret, and then a tiny golden flower beetle appeared, pacing back and forth on a nearby leaf. I had begun to think our quarry must be something nocturnal when a dark shape suddenly streaked directly in front of the lens. It was gone in an instant, but seconds later it reappeared and I paused the frame: there, fixed in mid-flight, was a perfect little black wasp. Thin-waisted and banded with creamy pale stripes, it hovered directly above the turret. But would it enter, or was it just passing by? Frame by frame we advanced the video as the wasp circled, like a helicopter approaching a landing pad in rough weather. Then it alighted on the rim, tucked its wings neatly across its back, and disappeared down the tunnel.

We gave a cheer of triumph, but I quickly realized that the video raised as many questions as it answered. After all, knowing that our quarry was a wasp didn't exactly narrow things down.

According to the *Catalog of Hymenoptera in America North of Mexico* (a 16-pound [7.2-kilogram] reference set in my office), more than 12,500 different species could claim that title, from solitary hunters to parasitoids, gall makers, hornets, yellowjackets, mud-daubers, and more.[10] So the next afternoon found me right back in the yard, crouched down beside the turret. The wasp's behavior followed a predictable pattern, darting off and staying away for ten to fifteen minutes, and then returning, circling, and re-entering the nest, where it stayed put for anywhere from a few seconds to half an hour. It moved so quickly I could make out little more than stripes and wing-blur until I used the camera to slow the action down. Then I saw exactly what the wasp was doing. Empty-handed on every outbound flight, it always returned bearing the same cargo gripped firmly in its mandibles: a small, lime-green grub. That meant the wasp was a hunter, and a female, bringing provisions to supply the brood she was tending in a tunnel belowground. (Male wasps don't hunt because they lack the necessary equipment—the stingers used to immobilize prey evolved from "ovipositors," egg-laying structures exclusive to the female anatomy.) Armed with that knowledge and a good picture, I soon identified our turret builder as a member of a group aptly named for their skillful use of clay and mud: the potter wasps.

Also known as masons or Eumenins, potter wasps occur on every continent outside Antarctica, enriching backyards and wild places with a wide range of adobe constructions. Jean-Henri Fabre had once found a species in his neighborhood building "spherical skull-caps" of mud attached to a garden wall, and there are others that fix their nests to twigs, tree trunks, boulders, lampposts, or leaves.[11] Some varieties prefer digging into earthen embankments, or plastering over the holes left behind by wood-boring beetles. Turret builders are relatively rare, which led me quickly to the genus *Odynerus*. Ironically, the name comes from a Greek phrase meaning "to cause pain," but potter wasps rarely sting. Whatever measure of belligerence they may possess is reserved for the caterpillars and other larval

FIGURE 2.2. A potter wasp in the genus *Odynerus* approaches its nest turret, carrying a snout beetle larva to feed to its young. Image © Thor Hanson.

insects they hunt. These they attack with surgical precision, injecting just the right amount of venom in just the right places to cripple, but not kill, their chosen victims. It amounts to a strategy for food preservation. Kept alive, their prey do not decompose and can be stacked up like cordwood in the nest—a living larder for their offspring to gnaw upon during the slow, dark weeks of their growth and development.

I knew that some version of this macabre scene was playing out just a few inches beneath my feet, but seeing it in action would have required digging up and destroying the nest. Fortunately, Fabre and others had done that work for me. Their efforts showed that *Odynerus* wasps liked to dine on the larvae of snout beetles—thus the female's single-minded focus on those little green grubs. She would need roughly twenty of them to provision each of the half-dozen egg chambers in her nest, an exhausting task that would dominate the remaining days of her adult life. The next time I returned to check on the turret it was gone. In a final maternal act, she had concealed her brood from the world, dismantling the nest's carefully sculpted entrance and using that soil to fill in the hole. All that remained was a granular, dime-sized scuff mark of slightly darkened earth.

Even when you know where to look, spotting evidence of potter wasps takes some luck. They spend the vast majority of their lives hidden away as larvae and pupae, emerging for only a few short weeks in spring and summer to mate, build nests, and start the whole cycle over again. Noah's timing had been perfect, and he certainly had that youthful knack for nonselective observation. But after crouching beside the turret long enough to get a leg cramp, it occurred to me that children benefit from another advantage when searching for things in nature: they're closer to the ground. Proximity matters when the cues are small, and while I might never recapture the full unfiltered potential of my senses, there was nothing stopping me from lowering my center of gravity.

The dictionary on my office bookshelf defines crawling as moving on hands and knees or dragging one's body across the ground. Either way, it promised to bring me face-to-face with a range of backyard dramas that, like the lives of potter wasps, had been unfolding unnoticed underfoot. But as soon as I dropped to all fours I realized something else. It made our backyard larger. Brief experimentation with a measuring tape drove the point home. When I walked upright at an average pace, I crossed 100 feet (30 meters) of open ground in 24 seconds. Crawling required 95 seconds to travel the same distance, effectively quadrupling the size of our property. Psychologists call this sort of change in perspective "the scaling of apparent distance." Jonathan Swift used it to good effect in *Gulliver's Travels*, where the ground-level viewpoint of the Lilliputians swelled their homeland—an island less than four miles across—into a grand domain stretching "to the Extremities of the Globe."[12] Although our yard didn't grow quite that large, crawling did transform the quick stroll of my daily commute into a substantial expedition.

Oddly, the first thing I noticed on that familiar trail had nothing to do with size. It was a smell. Facedown, inching through the ferns and shrubs, my nose suddenly filled with the pungent odor of a red fox. I knew that foxes passed through our yard, and I knew that they used urine and scent glands to communicate. Much like domestic dogs at fire hydrants, wild foxes deposit fragrant calling cards at key locations to send a variety of territorial and other social messages. But while I had often caught a whiff of foxy redolence in our woods, nothing prepared me for encountering a scent mark at nose level. Its muskiness was overpowering, almost eye-watering—so rich and feral that it was easy to imagine the subtleties of dominance or sexuality it might convey. I couldn't read the signal as a fox would, but something in me definitely recognized the basic text. I've rarely felt more like a mammal.

As I settled in to the low and lurching gait of my new mode of travel, my sense of hearing also seemed suddenly more intense. I noticed the background drone of insects mixed with the whine of a jetliner passing far overhead. My kneepads—a nod to middle-aged joints—scraped loudly across pebbles and sticks in the path, and a monotone beeping marked the progress of a delivery truck backing down the neighbor's driveway. Soon, however, those noises began to fade from my attention as the yard's minutiae came into clear visual focus. I saw mushrooms the size of toothpicks, and noticed for the first time how the tiny hairs on the stems of pathfinder, a native wildflower, were tipped with droplets of dark, purple fluid. It felt sticky to the touch, a natural flypaper to help defend the plant from the attacks of hungry insects.[13] Various castoffs from the world above carpeted the ground—fir needles, salal leaves, old feathers and cones, a tuft of deer fur. Looking ahead, I came face-to-face with the caterpillar of a silver-spotted tiger moth, all spiky with orange and black hairs. It sat immobile at the very tip of a young bracken fern, its tiny weight heavy enough to droop the frond groundward over the trail. The caterpillar was not feeding—was this where it would spin a cocoon and transform itself into an adult? I made a mental note to check on it again the next time I crawled to the office.

New perspectives invite new questions, and as I moved slowly along the path, glancing this way and that, one thought sprang repeatedly to mind: *How had I failed to notice all the spiderwebs?* At ground level, everywhere I looked seemed to shimmer with arachnid handiwork. Single strands and cables stretched among the grass stems; airy tangles filled the spaces between twigs. There were classic cobwebs and cottony domes, as well as tightly woven mats strung horizontally, like the safety nets deployed beneath high-wire acts. From an ecological standpoint, the presence of so many predators suggested an abundance of prey, and when I turned my attention to the air around me, I saw theory made plain in practice. Tiny insects abounded, the clear windows of their wings backlit with sunlight, their diversity conspicuous in the patterns of their flight. Some traversed a narrow arc back and forth as if attached to the arm of a metronome. Others rose and fell with the steadiness of pistons. I saw zigzaggers and loopers, and something with a reddish tint that dropped like a parachutist, drifting slowly past the others all the way down to the ground. It was mesmerizing, a display of sheer profusion that begged another question: Just how many different sorts of small things were living out their lives in our yard, unnoticed and unappreciated?

"One thousand, one hundred and eighty-one," Doug Tallamy told me. "And that's just moths." A career entomologist at the University of Delaware, Tallamy has been counting moth species in his yard for more than twenty years. "What's interesting to me is that the number is still growing," he said, and explained that even though his latest observations had been cut short by summer travels, he had still added thirty more varieties to his list. "We'll have well over twelve hundred by the time we're done," he predicted.

Tallamy lives in a white farmhouse on ten acres of woodlands and abandoned pasture in rural Pennsylvania. He chose to inventory moths on the property because, as he put it, "They transfer more energy from plants to animals than any other creature we know of." In other words, moths are foundational to food webs, devouring leaves and fruits as caterpillars, and then in turn being devoured by everything from other insects to spiders, bats, birds, reptiles, and mammals. A typical pair of chickadees, for example, must catch and deliver between six thousand and nine thousand moth caterpillars to raise up a single brood of nestlings.[14] Statistics like that make moths a good indicator of backyard biodiversity, something Doug Tallamy spends a lot of time thinking about.[15] His books and research on the vital connections between native plants, insects, and wildlife have inspired a grassroots restoration movement helping thousands of like-minded landowners convert their lawns into productive wildlife habitat. It was all a matter of rebuilding the food web. But today I had called him with a more straightforward question in mind. Humbled by my own limited familiarity with tiny, flitting things, I wanted to know how an expert reckoned with insect diversity. Even a single group like moths seemed a daunting thing to tackle. Just what did it take to get a better handle on that teeming cloud?

"I use a sheet and a light," Tallamy said simply. He went on to explain that while the occasional specimen turned up trapped inside his house or garage, he found the vast majority of his moths by co-opting the same navigational confusion that draws them to porch lights.[16] An ordinary bedsheet, brightly lit and hung outside in the dark, created a pane of white light that attracted night fliers in droves. It also gave them a convenient place to land, allowing Tallamy to take close-up photographs of their patterned wings and bodies before shutting down the lights and letting them go. All he had to do then was match up his moth pictures with the images available in books and various online identification guides. Anyone could do it, he said. In fact, anyone has.

Once a technique known only to specialists, "lightsheeting" has become a powerful tool for all sorts of backyard moth enthusiasts. Over the past decade, National Moth Week has grown from a single group of volunteers huddled around a sheet in New Jersey to an annual moth identification blitz spanning all fifty states and eighty countries. Similar efforts in the United Kingdom now continue year-round, contributing millions of citizen-generated records used for studying everything from moth abundance and diversity to population trends, the effects of urbanization, and the spread of introduced species. The lightsheeting method appeals to scientists because it is simple and repeatable, and the photographs provide reliable data. For everyone else, it opens a rare window into the hidden world of small creatures living all around us. And the public appetite for that kind of knowledge can sometimes be surprising. When a professor at San Francisco State University suggested a daytime

FIGURE 2.3. The practice of "lightsheeting" attracts a wide variety of nocturnal fliers, including the white ermine moth (*Spilosoma lubricipeda*) pictured in the foreground. Illustration © Chris Shields.

equivalent—planting sunflowers to attract and study backyard bees—the response nearly crashed her website.

"We thought we'd been hit by a spammer," Gretchen LeBuhn told me with a laugh, recalling the week her Great Sunflower Project went live. She had known there would be some level of interest—several groups of master gardeners had already been in touch about signing up. But nothing prepared her for the deluge of clicks and contacts that followed. She had to cut short a long-planned vacation and rush back to the office, watching in astonishment as the number of people eager to participate shot past fifteen thousand in a matter of days. That total continued to climb, and LeBuhn now coordinates a network of more than one hundred thousand dedicated backyard sunflower watchers across North America.

The concept is simple. Every participant plants the same Lemon Queen variety sunflower—a repeatable standard—and agrees to take notes on the bees that come to visit. That alone accomplishes the first goal of the project: getting more

FIGURE 2.4. Volunteers track backyard bee activity across North America by observing the flower visitors attracted to Lemon Queen sunflowers (*Helianthus annuus*). Illustration © Chris Shields.

people outside learning about pollinators. But the details of their observations matter too, because the project has always been about more than education alone. "I'm a data person," LeBuhn reminded me, and explained that her intent from the beginning had been scientific, to learn things about bees that couldn't be learned in any other way. Only an army of volunteers could work at such a vast geographic scale, uncovering regional trends that would otherwise go unnoticed. "I want everything to be rock solid before we publish," LeBuhn said, with a scientist's characteristic caution. She plans to amass several more years of data before releasing the first analysis, but preliminary results look promising. Variations in the number of bees on backyard sunflowers have already identified clear pollination hot spots and cool spots across the continent, along with something sure to add fire to the debate about insect declines in agricultural areas. Wherever pesticides or pesticide-treated seeds are commonly used—specifically those in the controversial neonicotinoid class—bee activity drops off sharply. "This decrease is not simply due to differences in land use or climate," reads a note on the project's website, accompanied by a plea for more observations to confirm the trend.

When I asked LeBuhn why the Great Sunflower Project was so popular, her answer echoed the combination of curiosity and concern that underlies all backyard biology. "People hear about pollinator declines in the news and they want to help," she said, adding that in an era of so many large-scale environmental problems—from insect declines to habitat loss to climate change—there was a need for simple, tangible responses. But anyone pitching in to efforts like LeBuhn's gains something more: a practical insight into the workings of science, and how one question inevitably leads to more. The project has already expanded to include other flowers, tracking how bees use different garden plants at different rates in different parts of the country. And then there are all the other small creatures showing up in people's counts. What about the flies and butterflies, the wasps, beetles, ants, and spiders? For that matter, what about the different kinds of bees? Unlike moths on a light sheet, bees rarely sit still for photographs, so LeBuhn's volunteers aren't required to identify them beyond broad categories. (LeBuhn analyzes the number of flower visits as a general indicator of bee activity and abundance.) It's probably just as well. Knowing one bee from another often requires pinned specimens and a microscope, or even a gene sequencer. And some species can't be identified at all, because nobody knows what they are.

One of the most surprising aspects of backyard biology isn't the number of small things living alongside us, or the number of them we've never noticed before. It's the number of them that *no one* has ever noticed, including scientists. Doug Tallamy was emphatic when I asked if—as a trained entomologist with decades of experience—he had ever encountered a moth he couldn't put a name to.

"Many!" he said without hesitation, and I could imagine him shuffling through the cabinet of unidentified specimens in his mind. "Many, many," he repeated. "It's not hard at all to get one."

Taxonomists, the scientists charged with identifying and cataloging new species, put the number of unknown and undescribed moth varieties in North America at more than two thousand. And moths are considered well studied. Few would venture to guess at the rich diversity hidden away in lesser-known groups like fungus gnats, springtails, leafhoppers, or midges. In the twenty-first century, an age of information overload, it can be shocking to learn that most small things on this planet (and quite a few larger ones) have yet to be named, let alone studied in detail. "We don't know what's out there," Tallamy stressed, and by "out there" he wasn't referring to some distant, unexplored wilderness, or even to relatively large, rural yards like his or mine. Virtually any patch of habitat contains something little known or mysterious, even in one of the most densely settled urban landscapes on the planet.

PART TWO

Exploring

To the lover, especially of birds,
insects, and plants, the smallest
area around a well-chosen home
will furnish sufficient material
to satisfy all thirst of knowledge
through the longest life.

—Mary Treat
Home Studies in Nature (1885)

CHAPTER THREE

Something New

There is nothing like looking, if you want to find something.

—J. R. R. Tolkien
The Hobbit (1937)

It all started with a wager at a lunch party.

"I had been bragging, perhaps unwisely, that I could find a new species anywhere," Brian Brown told me over a Zoom call. His dark hair and boyish face belied more than thirty years' employment at the Natural History Museum of Los Angeles County, where he currently serves as curator of entomology. He made that fateful "new species" boast at a donor luncheon, and one of the museum's trustees immediately took up the challenge. If Brown could find a new species of insect in her backyard, she told him, she would host a grand dinner party in celebration. To add a bit of pressure, she went ahead and scheduled the party.

"I didn't manage to get the trap set up until three weeks before the dinner," Brown recalled, but any angst he felt about the situation disappeared as soon as he brought the first batch of specimens back to his lab. With a practiced eye he began sorting out small, dark flies in the family Phoridae, his specialty. "The very first one I looked at didn't key out to anything," he said, referring to the technical manuals, or "keys," used in taxonomy to tell one species from another. To someone of Brown's experience and expertise, that could only mean one thing.[1] "It was unknown!" he said, still visibly excited years after the fact. The second fly he examined had "a weird leg," and turned out to be the first North American record of a species known only from Europe. Then he found one from a genus never before recorded west of the Rocky Mountains. "I didn't even have to finish looking through the sample," Brown went on. "I already had enough for the party." More than that, he had confirmation that backyards in cities were worthy of exploration. And that gave him the beginnings of an ambitious new research program.

"The BioSCAN project is unique," Brown explained, ticking off the features of an effort now in its eleventh season. Where most biodiversity surveys explore rural places and wildlands, BioSCAN is focused on greater Los Angeles. And where most surveys inventory a wide range of life-forms, Brown's group decided to "go narrow and deep," studying specific insects like phorid flies and bees. They've sampled from the urban core to the edge of the city, and from the coast inland to the desert, and they have consistently avoided designated natural areas, targeting instead the sorts of places usually overlooked by biologists—backyards, gardens, and small neighborhood parks. The results have been no less remarkable than those first few flies from the trap in his trustee's yard. "There's no end to it!" Brown exclaimed. "Every time we sample in a new area we get new species." For phorid flies alone, the team has discovered fifty species new to science, and expanded the known ranges of scores of others, and they still have thousands of specimens left to examine.

To Brian Brown, the success of BioSCAN's backyard strategy is gratifying, but not surprising. He finds similar untapped

FIGURE 3.1. The BioSCAN project has discovered dozens of new insect species in typical backyards like this one in greater Los Angeles. Here, the trap sits on a patch of lawn grass, squeezed between a hedge and a hot tub. Image © Natural History Museum of Los Angeles County.

diversity wherever he looks for flies. Phorids are his forte, but the entire fly order, Diptera, remains poorly known. Over 160,000 species have been described, with perhaps ten times that number waiting to be discovered. Or maybe more—nobody knows for sure. As Brown put it, "The magnitude of the richness, and of our ignorance, is beyond what we're used to thinking about." Like crawling through the yard to recalibrate our senses, overcoming our ignorance of flies requires a purposeful change in perspective.

Part of the problem boils down to public relations. In spite of the fact that far more fly species are attracted to flowers than to rotting meat, we tend to associate the entire group with filth and decay. Phorids have a particularly bad time of it. First of all, "phorid" rhymes with "horrid," an unfortunate phonetic echo of our worst preconceptions. Common names for the family are no better: hunchback flies, for the distinctive bulge in their thorax; scuttle flies, for their habit of running jerkily across surfaces; and coffin flies, for the *single* species of phorid that does, admittedly, like to hang out around cadavers. Giving this image a makeover might sound like a daunting task; but if anyone is up to the challenge, that person is Emily Hartop.

"Phorids do everything!" Hartop gushed, when I asked why the family held so many species. It occurred to me that the same might be said of Hartop. I reached her in Poland, where she was working on a side project five days before defending her doctoral dissertation in Sweden, a task she planned to complete en route to Berlin, where she was soon beginning a new job at Germany's foremost natural history museum. All of this while still publishing discoveries from her years working alongside Brian

Brown on the BioSCAN project. That's where she first became interested in phorids, a group she now speaks about with a mixture of calm passion and certainty, as if it were a foregone conclusion that everyone would share her love for these tiny flies just as soon as they got to know them.

"Charismatic insects get all of the attention," Hartop said, dismissing the showy butterflies and bees of the world with a shrug, "but they don't have nearly the same diversity of life histories." She has a point. Among insects, most groups have perfected playing a single role in an ecosystem. Potter wasps are good at catching grubs, for example, and termites specialize in eating deadwood and other decaying plants. But phorids are like ecological polymaths. They are "hyper-diverse," as Hartop put it, with species that have evolved to fill every imaginable insect niche, from predators and parasites to scavengers, herbivores, fungus eaters, decomposers, and more. There are phorid flies whose larvae eat only frog eggs, and others that prey upon the offspring of particular beetles, or the crushed bodies of certain snails or millipedes. Some phorids live inside ant nests or beehives, stealing food and attacking weak larvae. There are phorids that burrow underground, grazing on the mycelia of mushrooms, and others that live in cornfields, feeding on dead leaves. Some phorids pollinate tropical rainforest trees; others visit European buttercups or the flowers of arctic willows. In certain cases, the lifestyles of phorid flies border on the downright bizarre. One of the world's smallest insects is a phorid, for example, with a body that measures only 0.016 inches (0.40 millimeters) in length, so minute that it experiences air less as a gas than as a liquid, using its tiny wings like paddles to push itself through a viscous sky. And then there are the ant-decapitating phorids, living up to their name by developing *inside* the heads of live ants, where they feed on brain tissue and eventually produce an enzyme that cleaves head from thorax as neatly as a guillotine.[2]

Speaking to Hartop made me increasingly curious about what sorts of phorids might be living in our yard. Apparently, she has that effect on everyone. "People get really excited," she admitted, describing how BioSCAN participants reacted to the backyard flies and other insects that she and her colleagues showed them. "When they look outside they see grass, trees, maybe a swimming pool, or their kids' toys," she said, "but they don't have any idea what lives there." Realizing that scores or even hundreds of different species share that habitat, and that some of them might be unknown to science, quickly changes people's perspective. Lisa Gonzalez, another BioSCAN scientist, told me how her monthly visits to check on participants' traps often turned into impromptu entomology lessons for the whole family. Occasionally, people's newfound enthusiasm got carried away. She recalled how one trap seemed to catch an inordinate number of isopods, the little armored crawlers commonly called pill bugs or roly-polies. Month after month the numbers were off the chart, until the family's six-year-old daughter proudly announced that she had been making her own backyard bug collections, scooping up handfuls of roly-polies and dropping them into the trap to be tallied.

I felt a similar urge to participate, but BioSCAN's efforts don't yet extend beyond Southern California. Still, I have a microscope and a basic knowledge of insects, so I asked Hartop if she thought I could replicate the project—albeit on a small scale—in my own yard. "I can help with that," she answered immediately, taking on another commitment with the same unselfish drive that has led to such accomplishments so early in her career. (Most scientists don't start discovering species, publishing dozens of papers, and collaborating in multiple countries until *after* they finish their PhDs.) Soon I found myself at the edge of our woods in a sunny, south-facing nook, unfolding a tentlike structure of nylon mesh disconcertingly known as a Malaise trap. Although meant as an homage to its Swedish inventor, René Malaise, the name evokes the malaise of an uneasy conundrum: how the study of insects so often involves killing them.

In ornithology and zoology, lethal sampling tools like shotguns have largely given way to less invasive techniques, such as remote cameras, satellite trackers, or a good pair of binoculars. Less so in entomology, where killing jars still play a vital and unavoidable role. The difference lies in the scale of the subject matter. Unlike birds and mammals, few insects can be productively studied from a distance. Most can't even be identified. Understanding anything about their lives requires taking specimens, so entomologists have learned to become opportunistic and thrifty. Hartop inspects the grilles of cars after road trips, for example, and the trap in my yard presented a similar chance for unexpected data. By sampling a few square feet (roughly one square meter) of airspace for a few days every month for a year, the Malaise trap promised a low-impact snapshot of seasonal insect diversity—catching just a few individuals of whatever happened to be flying around the yard. I agreed to sort out the phorids and send them to Hartop, and it didn't take long to find other entomologists eager for the bycatch. A friend at the Department of Agriculture wanted sawflies (a group that includes many crop pests), for example, and taxonomists at the American Museum of Natural History asked for all the bees, wasps, and ants. Collectively, their curiosity amplified my own, and as the days passed I could hardly wait to see what the trap had collected.

At first glance, my insects didn't look like much. The trap funneled everything from the mesh screen upward into a small jar filled with ethanol. It had started out clear, but now resembled a brownish sludge. Still, Hartop had warned me not to pass judgment until I put things under a microscope. That turned out to be good advice. Back in the Raccoon Shack, I whispered an involuntary "*Wow*," as the jumble of murky shapes came into abrupt focus. Gold and metallic green glinted from the wing scales of an owlet moth, and the tiny thorax of a cuckoo wasp shimmered with iridescence. Red-eyed fruit flies, gnats, and small yellow hoppers crowded amongst the gangly legs of a crane fly, and two knobbly snout beetles framed the slender emerald body of a lacewing. And that was just the beginning. Scores of species filled the frame, a dizzying assortment of angular forms, most of them wildly unfamiliar, like spare parts in an alien junkyard. Encountering abundance in nature is always exhilarating—there is a reason why tourists and documentary filmmakers (and biologists) flock to witness spectacles like the herds of the Serengeti. But finding profusion so close to home, in a group of creatures so underappreciated, surprised me emotionally as well as intellectually. Magnified and celebrated, our backyard clouds of tiny fliers are wondrous—movingly so. I suddenly understood something that Brian Brown had told me: experiencing the hidden diversity of their yards had brought more than one BioSCAN landowner to tears.

Entomologist E. O. Wilson famously called invertebrates "the little things that run the world," and phorid flies are small even in that diminutive company.[3] But when I zoomed in, the microscope revealed their compact bodies in clear and bristly detail. Up close, they looked like the insect equivalent of weightlifters, with tiny heads perched on hulking thoraxes, and overdeveloped back legs that splayed out behind as if too muscular to fold beneath their abdomens. Even to the uninitiated, their diversity was obvious. They ranged in color from jet-black to brown, amber, or gray—sometimes unpatterned, and sometimes boldly striped with white. There were species adorned with spikelike hairs on their heads and backs, while others were smooth and shiny, and they ranged in size by at least an order of magnitude, from the merely minute to, well, microscopic. One feature—or lack thereof—seemed to unite the whole family. Where the wings of most flying insects are patterned by a gridwork of dark veins (think stained glass without the colors), the phorids boasted simple, transparent membranes lined with only a few graceful curves, like wing shapes sketched by an impressionist. (Brian Brown later complimented me on noticing this trait—a good identifier, he said, although, in yet another example of phorid diversity, it wouldn't help me sort out the hundreds of species that lacked wings altogether!) As I began setting the phorids aside, they quickly returned to anonymity—just a jumble of small black flecks floating in a vial. But every

FIGURE 3.2. Members of the hyper-diverse phorid fly family display a huge range of life histories, from predators to plant eaters, fungus eaters, scavengers, wood borers, leaf borers, ant-decapitating parasitoids, and many more. (Clockwise from upper left: *Apocephalus* sp., *Megaselia marquezi*, *Megaselia scalaris*, *Dohrinphora cornuta*). Illustration © Chris Shields.

time I spotted another variety I couldn't help wondering whether mine weren't the first human eyes to pause upon it, however briefly, as something distinct and unique.

In an era of high-tech laboratories, sorting through insect specimens with a pair of tweezers can feel like an activity from a different era. But it's a powerful reminder of how little we know—not just as individuals, but collectively. Discovering new species remains fundamental to the study of biology, and the folks at BioSCAN are far from the only experts to recognize the potential of backyard collaborations.[4] Public participation is essential in addressing what scientists call the "taxonomic impediment," the great imbalance between the vast number of species yet to be identified (estimated at more than 80 percent of the Earth's total) and the relatively few professionals trained to do the job. Yes, the final steps in determining new species require expertise, but anyone can gather the data and make the observations that get the process started. In Europe, for example, amateur naturalists contributed to 62 percent of all new species descriptions published from 1998 to 2007, including insects, spiders, worms, reptiles, and amphibians. That pattern repeats itself wherever taxonomists can find willing partners.[5] Teaming up with local mycology clubs has unearthed a flood of new mushroom species from the southern Rocky Mountains, while appeals to recreational scuba divers and beachgoers have revealed new jewel anemones in Singapore, new bristle worms in the Caribbean, and a new nudibranch in Norway. Most projects involve habitats in and around where people live, such as the new wasp found during a volunteer survey of Amsterdam's busiest urban park. But species discovery can also happen on vacation. Participants in taxonomy-themed ecotours have recently chalked up new beetles and a semi-slug in Borneo, and several new water mites, slugs, and snails in Montenegro. (Trips by companies like Taxon Expeditions regularly sell out, attracting customers/donors with the slogan "You can be Darwin too!") Opportunities in educational settings have also borne fruit, including one that boded well for my own efforts. When students deployed Malaise traps on the grounds of four primary schools in South Australia, and tended them in just the way I was doing, every trap yielded at least one new species of insect.[6]

Species discovery often accelerates when experts and backyard enthusiasts work together, but the fact of the taxonomic impediment remains. There simply aren't enough trained specialists to go around, and most people won't have one at their elbow if and when they stumble across something new. Fortunately, there is now an alternative, a device that nearly everyone in the modern era keeps close to hand at all times, conveniently embedded in their smartphone. Perhaps more than any other recent innovation, the digital camera is transforming how people learn about nature. But it's also transforming how scientists learn. Which is precisely what ecologist Scott Loarie had in mind over ten years ago, when he left academia to become co-director of a quirky little website called iNaturalist.

"I wanted to do something that wasn't just a paper," Loarie recalled, and described the frustrating gap between researching a problem and actually doing something about it. As a postdoctoral fellow at Stanford University, he had published a major study in 2009 on "the velocity of climate change," the idea that climatic conditions were sweeping toward the poles and up in elevation at a predictable rate as the planet warmed. Animals and plants would have to shift their ranges at the same speed if they wanted to stay within the environmental parameters they were used to. Loarie's model had important implications for biodiversity, because it implied that many species would soon be moving away from the parks and conservation areas created to protect them. Or if they stayed put, they might not survive as their habitat morphed around them into something quite different. "The paper got a lot of press attention," Loarie told me. "People kept asking 'What's next? What's next? How do we address the problem?'" But he found that he had little to tell them. There simply weren't enough real-time data on the locations of plants and animals. Were species moving? If so, which ones and how fast? Nobody knew.

It was about that time that Loarie met Ken-ichi Ueda, a graduate student who had codeveloped iNaturalist as a project for his master's degree.[7] The concept was to create something like a social network for naturalists, an online community where people could share photos and expertise, helping one another learn about the plants and animals in their neighborhoods. Ueda had a vested interest in the outcome. Recently arrived in California, and an avid naturalist (he has described himself as "totally incapable of ignoring nature"[8]), Ueda was eager to get his head around all of the unfamiliar species in his new home. He saw iNaturalist as the perfect tool to do so—and if it worked for him, it should work for anyone. Scott Loarie agreed, but he saw something more. Every photograph uploaded to the site represented a discrete observation of an individual organism, in a particular place, at a particular time. Verify the species and get enough people sharing images, and iNaturalist could provide just the sort of data that he and other climate scientists had been longing for. Loarie told me that it was an easy decision to switch gears and start working with Ueda full-time. "I knew that this was exactly what I wanted to be doing," he said. What he didn't know—at least not right away—was how the platform would explode in popularity, connecting naturalists across the globe and generating one of the largest biological datasets in history.

"We now have 120 million observations of more than 400,000 species," Loarie told me. "That's one out of every four named species in the world."[9] He sounded a little breathless as he cited those statistics, and then explained that he was pressed for time before another commitment, and hoped I didn't mind if he took a brisk walk to get some exercise while we talked. That sort of double-tasking is probably routine at iNaturalist, where enthusiasm for nature and science blends freely with the caffeinated energy of a tech startup. If it were a commercial enterprise, investors would drool at their growth charts.[10] The number of both users and observations has roughly doubled every year since 2015, producing a collection of images from more than 180 countries on a platform now available in eighty-three languages. "That community is what makes iNaturalist most unique," Loarie said, and stressed that the emphasis remains on bringing people together—maintaining a friendly online space for the sharing and discussion of natural history observations. "The data are a byproduct," he told me, but then excitedly acknowledged that scientific use of those data has been growing at the same incredible rate.

"Over three thousand papers have used iNaturalist data," Loarie declared, and then he explained how the process worked. To be deemed "research grade," images uploaded to iNaturalist needed to be verified to species and accompanied by GPS coordinates and a date stamp. (Smartphones and cameras often include the latter information automatically.) Observations that met that standard were then passed along to the Global Biodiversity Information Facility (GBIF), an international clearinghouse for "open access to data about all types of life on Earth."[11] Founded in 2001 and funded by more than forty governments, GBIF began as a digitized catalog of specimens housed at natural history museums and universities, allowing scientists to study them online rather than having to travel laboriously (and expensively) to every collection of interest.[12] Those records, however, are now far outnumbered by contributions from citizen science projects, notably iNaturalist and eBird, a popular bird-watching platform managed by the Cornell Lab of Ornithology. Loarie lauded this trend as "the democratization of science" and then made a comment that surprised me: "A lot of scientists don't even know they are using iNaturalist data." He explained that researchers regard GBIF as a trusted source, so they don't necessarily scrutinize the details of every piece of information they download. "It's a little bittersweet," Loarie mused. He was glad that the data were making a contribution, but sorry that iNaturalist, and all of us who use it, weren't getting more credit. Still, that omission may stand for something far more important: a tacit acknowledgment that observations made by nonprofessionals—often in their own backyards and neighborhoods—don't need special treatment. They have become part of the scientific firmament.

Before the end of our conversation, I pushed back a little on Loarie's zeal for new technologies. He had told me that machine learning from the millions of images in its database now allowed iNaturalist to identify many species automatically, without needing to consult the online community. But didn't that risk taking human learning out of the system? I asked him. If smartphones could recognize every species instantly, why would people bother committing anything to memory? Loarie agreed that nature-watching shouldn't become yet another activity dependent upon microchips and glass. "Screen time is a concern," he admitted. "I have kids—we talk about it!" But in a society already too plugged in, he felt that any way iNaturalist could encourage more people to get outside counted as a win. And Loarie also believed that nature knowledge was a distinct form of learning. "It's not like quantum mechanics or something," he told me. "That stuff is hard to think about because we're not really meant to be thinking about it!" Our brains are much better prepared to absorb natural history. Recognizing the different plants and animals in our environment was once a fundamental part of the human experience—an evolutionary survival skill honed over the thousands of generations our ancestors spent as hunters and gatherers. "We've been severed from that, but we're still really good at it," he said. "It's an unflexed muscle that people want to flex."

Judging from the exponential growth at iNaturalist, and all the scientific discoveries those observations are fueling, Scott Loarie may be onto something. After I hung up the phone, I spent some time wading through the sea of research papers that have cited iNaturalist data and found that they were just as impressive in their variety as in their quantity. Yes, there were studies tracking climate-driven range shifts, just as Loarie had expected. Photos of dune grasses in Virginia had helped document a northward surge along the Atlantic coast, while snapshots of moths from the hills around Hong Kong had found them moving upslope to cooler elevations. In California, dozens of sightings of juvenile great white sharks showed how they were following warm ocean currents from the subtropics all the way north to Monterey Bay. But climate change studies hardly scratched the surface. There were papers investigating everything from roadkill trends in South Korea to the spread of pet iguanas released into the parks of Florida. More than one effort examined the finer points of evolution, particularly color patterns and other traits easily identifiable in photographs. The dark patches on the wings of male dragonflies, for example, correspond to the temperature of their environment—larger in cooler places where absorbing warmth is an advantage, and smaller in hotter climates, where soaking up too much sunlight increases the risk of overheating. It took careful scrutiny of 2,718 iNaturalist images to reveal that particular pattern, but some discoveries require just a few pictures, or only one. Especially when people upload photographs of things that no one has ever seen before.[13]

In 2019, two San Francisco–area teenagers were scrolling through the platform when they noticed a scorpion photo that had languished unidentified for six years. Then they found another one. By devoting their school holidays to visiting the nearby desert lake beds where the pictures had been snapped, they were able to relocate and study the mystery creatures in the wild, adding two new species to the world's list of known scorpions.[14] Similar creativity with iNaturalist data has aided in the discovery of everything from raspy crickets to quillworts, assassin bugs, stiletto flies, and a narrow-leaved agave plant known from a single mountain range in Mexico. I particularly enjoyed one paper about a new banded rubber frog species from Angola, where the authors reprinted diagnostic photos that had obviously been taken on somebody's patio. That connection to home runs through all the work at iNaturalist and other citizen science platforms. As these online communities become more established, researchers aren't just scanning their pages for relevant observations. They have begun interacting with users directly, reaching out to people in particular places to search their neighborhoods for species or behaviors of interest. Speaking rhetorically, Scott Loarie put it this way: "Can we actually mobilize this huge group of amateurs to address pressing concerns in science and conservation?" The answer seems to be a resounding "yes."

My own backyard search for new species had come down to an email. I was waiting to hear from Emily Hartop's lab in Germany, where at least a portion of my phorid fly specimens had finally arrived. (To the list of challenges inherent in studying small insects, Hartop and I had added another: they get lost in the mail.) Using a combination of fine microscope work and genetic testing, she expected to process in a matter of hours what had taken me months of effort to accumulate. But when her message landed in my inbox, the news was disappointing.

"I am sorry I can't look at this in time for your deadline," Hartop wrote. In those few words, I came face-to-face with the taxonomic impediment. Specialists like Hartop always have a backlog of samples to work through—not just from their own projects, but from researchers all over the world eager to put names on puzzling, little-known organisms like phorids.

Sometimes it can take years to get an answer, but when so few people have the skills to do a job, there is really only one course of action left open: join the queue.

Eventually, my vials of flies will rise to the top of the pile, and when they do I have every reason to be hopeful.* Decades have passed since the last serious phorid study in my area, and when Brian Brown recently examined a few specimens from a neighboring county, one out of three appeared to be something undescribed.[15] But regardless of whether any of my backyard flies turn out to be new to science, they have all been new to me. I found myself repeatedly astonished not only by their abundance and diversity, but also by their hardiness. From summer droughts through fall storms and even on snowy winter days, when hardly another insect was visible, my trap always seemed to find a nice diversity of phorids on the wing. Which made me more curious than ever.

Taxonomic labels help us to recognize plants and animals, but that knowledge leads inevitably to an even larger question: What are they doing? Scott Loarie summarized the situation nicely: "A name is a password that unlocks learning." And learning about nature is not so different from learning about people, where introductions are just that—they come at the

* Days before this manuscript went to press, I received another email from Emily Hartop. She had run a preliminary genetic analysis that put the number of phorid varieties in my yard at 104. (And that was from only half of the specimens!) "Distinct clusters" in the data pointed to the possibility of new species, but confirming that would require dissections and other detailed microscope work. She promised to get to it soon and encouraged me to keep looking: "There may be considerably more to find."

beginning of a conversation. We should interact with plants and animals in the same manner, always remembering to prolong our attention beyond that initial moment of recognition. Doing so will quickly reveal that even the most familiar backyard creatures are complicated, fascinating, and very capable of behaving in ways that surprise us.

CHAPTER FOUR

Forces of Habit

As the occasion, so the behavior.

—Miguel de Cervantes
Don Quixote (1615)

That's a frog, I thought, my mind absently registering a familiar shape as I walked briskly along the road. Then the penny dropped. *A frog?!?* It wasn't the creature's identity that surprised me—Pacific tree frogs are commonplace in our neighborhood. What stopped me in my tracks was the frog's position: dangling by one back leg from the bill of an American robin.

I managed to focus my binoculars on the pair just as the robin took flight from its hedgerow perch, carrying the struggling amphibian high into the air and dropping it onto the roadway, barely 30 feet (10 meters) in front of me. Then the bird followed the frog down, pursuing it across the tarmac and lunging in with vicious pecks in a behavior that ornithologists call "bill pouncing." The limping frog made it as far as the grassy road verge before collapsing, and still the robin pressed its attack, striking and stabbing and repeatedly lifting the now lifeless body into the air and thrashing it against the ground.

Fans of theater or cinema experience a disorienting shock whenever well-known performers "play against type," acting out roles that run counter to their public image. I felt the same way watching the robin, a bird widely beloved across North America as a cheerful harbinger of spring. But here, instead of "Pour[ing] forth the gladness"[1] of its breast, the bird had opened up a can of predatory whoop-ass. I saw it land more than 150 blows on the frog: pretty strong stuff from a species better known for eating berries and tugging at the occasional earthworm. Later, I pored through ornithological references and confirmed that fruit and invertebrates dominate the diet of robins, but I also found sporadic reports of attacks on garter snakes, as well as salamanders, shrews, and a skink. In 1940, a biologist stationed near California's Mount Shasta made the "most unusual" observation of a robin stalking and catching small trout in a stream.[2] "For the sake of my reputation," he wrote about the episode, "I am happy that several reliable witnesses can be named."[3]

Although I couldn't call on any witnesses to corroborate my frog story, the robin left me with something even better: hard evidence. After six relentless minutes, the bird's wild pecking began to slow until it finally stopped and stood stock-still beside its prize, apparently exhausted. After a brief rest and a few

FIGURE 4.1. My walk through our neighborhood was interrupted by the unexpected sight of an American robin (*Turdus migratorius*) attacking a Pacific tree frog (*Pseudacris regilla*), carrying it high into the air, and dropping it onto the pavement. Illustration © Chris Shields.

more listless jabs, it gave up entirely and flew off, allowing me to collect the body of the poor frog. I immediately cut short my neighborhood walk and hurried home to the Raccoon Shack, eager to conduct a full postmortem.

By all appearances, the frog had died from what the coroner in a crime drama might call blunt-force trauma. Black welts crisscrossed its back, head, and legs wherever the robin had struck, but nowhere had the bird managed to pierce the frog's skin. Although clearly strong and nimble, the robin's bill lacked the hooked, tearing tip that habitual predators like hawks, owls, and shrikes use to open the bodies of their prey.[4] That fact alone may account for why robin attacks on vertebrates are rare. For all the energy the bird expended catching, dispatching, and attempting to dismember the frog, it received not a single calorie in compensation. Racking up that kind of negative balance flies in the face of optimal foraging theory, the idea that species succeed by maximizing their dietary efficiency, focusing on foods that bring the largest reward for the smallest investment of time and effort. With such high costs and no good meal in return, frog hunting made zero ecological sense for a robin. Yet that math obviously hadn't stopped this individual—a healthy-looking adult male with a lifetime of foraging experience. Had it hunted frogs or other small vertebrates successfully in the past? Had it fed on roadkill? And how had it learned to release the frog in midair directly above the pavement? Did robins deserve a place alongside corvids and gulls in the exclusive club of "prey droppers," those birds clever enough to intentionally exploit the combined effects of gravity and hard surfaces?

For a simple robin to spark so many new questions shows just how much remains to be learned about the habits of even the most common backyard species. (One recent estimate puts the total number of American robins at 370,000,000 individuals, which ranks it well above various sparrows, finches, and warblers as the most abundant native songbird on the continent.[5]) It also makes a powerful case for emphasizing the "watching" half of bird-watching. Too often our observations of birds—or any other wildlife—end at the moment of recognition. We look just long enough to see what something is, and then turn away, neglecting to ask the next logical (and arguably more interesting) question: What is it doing? To really understand what is happening in the natural world, we need to pay attention to behavior. That's not always easy. Watching closely takes time, a precious commodity that is often hard to spare, particularly for familiar species that our eyes tend to pass over as part of life's backdrop. But the robin and frog incident inspired me to try. Again and again, forcing myself to pause and watch common creatures has revealed intriguing behaviors. A hummingbird poking around the bark of a pine stump, for instance, turned out to be gathering spiderwebs to build its nest. And I once saw a woodpecker deliberately startle a saw-whet owl, landing directly above its target and tapping loose a dead branch that crashed downward, driving the owl from its roost.[6] Few places offer a more convenient stage for studying such dramas than the yards and byways closest to home. There is a long history of backyard contributions to the study of animal behavior. Think of Darwin and his earthworms at Down House, or Gregor Mendel studying honeybees (as well as pea plants) in the gardens of St. Thomas's Abbey. Nobel Prize winner Konrad Lorenz made so many discoveries about geese and jackdaws at his family's home in Altenberg that the Austrian Academy of Sciences declared it a research institute, paying Lorenz a stipend, sponsoring students, and funding the construction of aviaries, terraria, and a greenhouse. But not all famous examples occurred on the quiet grounds of abbeys or country estates. One groundbreaking discovery about insects took place at Haines Normal and Industrial Institute in the urban heart of Augusta, Georgia, when the resident high school science instructor decided to spend his morning off exploring an overgrown flower bed.

History does not tell us what Charles Henry Turner noticed first, the bee or the bottle cap. We do know that a "gentle breeze was blowing from the south"[7] on that sunny Saturday morning in 1908, when Turner made the following observation: "In a barren spot in this bed, adjacent to an inverted tin cap of a coca-cola bottle, and within an inch of the northern face of one of the bricks that formed the serrated border, a burrowing-bee had excavated a burrow."[8] For most of us, that might have been enough—it's not every day that one spots the nest of a secretive, solitary bee. But Turner saw something more. He immediately recognized the great potential of the bottle cap. Situated so prominently beside the entrance to the nest, it offered the opportunity to solve a nagging biological puzzle: How do bees find their way home?

In Turner's day, most experts regarded insects as "reflexive machines," with simple brains incapable of more than preprogrammed, instinctive responses. By that way of thinking, their homing abilities amounted to something innate, a mindless behavior triggered by the smell of the nest, the angle of the sun, or some other predictable environmental cue. Turner put this theory to the test with a simple and elegant experiment: he moved the bottle cap. While the bee was out foraging, Turner shifted the cap a few inches to the west and used a sharp stick to dig a false burrow, setting the stage for a telling drama. "The flowers from which it obtained its supply must have been quite remote," he wrote, "for it required about thirty minutes to make a trip."[9] But when the bee finally came back, Turner's patience was amply rewarded. He watched as it flew slowly along the edge of the flower bed until it reached the bottle cap. Once there, it landed immediately and entered the false nest.

Thus began a whole series of trials and manipulations involving more bottle caps, more fake burrows, tiny paper tents, and other challenges designed to show that bee navigation was anything but robotic. Turner's flower bed studies methodically demonstrated that bees have a sense of geography, recognizing topographic landmarks and using them to pinpoint the locations of their nest holes. What's more, his subjects readily adapted to changing conditions, rarely entering a false nest more than once, and quickly updating their mental maps when he added more bottle caps, sheets of paper, melon rinds, and other quirky features to their environment. Far from automatons, the bees were clearly capable of learning and remembering. "By a process of elimination," he concluded, "the most consistent explanation of the above behavior is the assumption that burrowing-bees utilize memory in finding the way home."[10] Over a century later, after scores of additional studies using increasingly sophisticated techniques, Turner's insight remains unchallenged.

For Charles Henry Turner, conducting research in schoolyards and other neighborhood settings wasn't just a convenience; it was a necessity. As a Black scientist in a deeply prejudiced age, he was largely shut out from research positions at major universities, and never had regular access to a laboratory, graduate assistants, or an academic library.[11] Sociologist W. E. B. Du Bois knew Turner personally, and pointed to his career as "a tragedy," a case study in potential thwarted by racism.[12] "He was a promising scientist," Du Bois wrote; "with even fair opportunity he ought to have accomplished much; but his color hindered him."[13] Limited to working in his spare time and without institutional support, Turner still managed to amass a remarkable list of achievements. He produced more than seventy scientific articles on topics ranging from decision-making in snakes to color vision in honeybees to the web-building habits of gallery spiders. Along the way, he became the first Black scholar to earn a doctoral degree in zoology from the University of Chicago, and the first to publish his research in the prestigious journal *Science*, a feat he accomplished three times.

The fact that Turner's studies took place in commonplace locations like flower beds and parks is a reminder that sharp eyes can spot intriguing behaviors anywhere, even in the middle of cities. That's precisely what behavioral ecologist Barbara Klump was counting on in 2018 when she began advertising on local radio stations in Sydney, Australia, recruiting people from throughout the metro area to answer the same unusual question: Had they ever seen a parrot opening the lid on their garbage bin?

"We were also interested to know if they hadn't seen it," Klump clarified, when I reached her on a Zoom call. "It was crucial that we also could say where the behavior *doesn't* happen." By including that key detail in her protocol, and by racking up reports from over 1,300 participants in 478 different suburbs, Klump was able to pinpoint the three places where the bin-opening technique had originated. It helped that the behavior was so distinctive—anyone would notice a parrot strewing garbage around their yard. It also helped that the bird was so distinctive—among Australia's fifty-seven parrot species, only one is large and snowy-white, with a comical plume of yellow feathers flopping around on its head.

"Sulphur-crested cockatoos are really ideal for this kind of study," Klump told me. Intelligent and highly social, the birds live in what she called a "fission-fusion" system, dispersing in small flocks to forage during the day (the fission), and then gathering together again at large communal roosts to spend the night (the fusion). "The foraging parties change constantly, giving them lots of opportunities to learn behaviors," she explained, and that learning process lay at the heart of Klump's research interests. After making a name for herself studying tool use in Hawaiian and New Caledonian crows, Klump embarked on the cockatoo study as part of a postdoctoral fellowship based at the Max Planck Institute of Animal Behavior in Germany. The project took shape quickly as soon as she and her colleagues documented the bin-opening behavior and showed that it was brand new, limited to just a few birds in a handful of locations.

"Oh my god!" she said, remembering her first glimpse of a cockatoo in action—how it used its bill to pry at the bin's lid, pushing upward and then sidestepping along the rim until the lid flipped back on its hinge with a bang. "All of these questions

FIGURE 4.2. Backyard observations helped track the spread of bin-opening behaviors learned by sulphur-crested cockatoos (*Cacatua galerita*) across forty-four neighborhoods in Sydney, Australia. Image © Barbara Klump/Max Planck Institute of Animal Behavior.

started coming in. How do they figure it out? Do they all do it in the same way? Is this a socially transmitted behavior?" Finding answers would require a lot of data, far more than she and a few student helpers could ever accumulate on their own. "We couldn't have done it without the citizen scientists," she said, and explained how she continued her online surveys for years, gathering observations from thousands of residents across greater Sydney. Some of them still keep in touch with her, sending pictures, notes, and videos of bin-opening and other cockatoo antics playing out in their neighborhoods. Those legions of volunteers were essential to the project, but it turns out that working in a backyard setting offered another important advantage.

"Studying animal behavior is usually really tricky, especially in the wild," Klump told me, and she listed a range of issues that often crop up to complicate a clear analysis—differences in food resources, vegetation structure, terrain, and other habitat variables that make it hard to compare behaviors from different places or times. The garbage bins of greater Sydney, in stark contrast, are completely standardized. Municipalities require everyone to use the same model: square and plastic, with a bright red top. Recycling bins feature yellow lids, a distinction the birds quickly learned to recognize. From the standpoint of study design, this gave Klump and her colleagues a level of control rarely experienced outside of a laboratory. As she put it, on every bin-day, all across the city, "Thousands of birds have access to the exact same resource."

Because Klump's team had thought to document both the presence and absence of bin-opening cockatoos, and because they had spotted the behavior so soon after it began, they found themselves in a position to map its spread. From those three original locations, the ability to open garbage bins swept through cockatoo populations in an additional forty-one suburbs in only two years. And as it diffused, the behavior began to change.

"It's a difficult task for the birds," Klump explained. "They try and try and often fail." That profusion of repeated attempts gave the cockatoos ample opportunity to experiment, and to develop variations on the original technique. Some grabbed the lid with their bills, others used a foot, or a foot and bill together. Some walked to the right along the rim; others went to the left. Some birds sidestepped, while others walked straight. Those nuances could be mapped too, and it quickly became apparent that the birds were doing more than just learning from one another; they were developing distinct neighborhood styles. And when a socially learned habit begins to vary geographically, taking on unique traits in different populations, behavioral ecologists have a word for it: culture.

"The situation is quite unique," Klump said, reflecting on the combination of setting, observations, and timing that had led to her team's discovery. Finding evidence of animal culture is always a rare event—variable tool use among chimpanzee populations, and distinctive feeding behaviors learned by certain whales, dolphins, and crows stand among the few unequivocal examples.[14] But Klump's team had borne witness to something even more unusual: the *development* of culture, watching a social trait spread and change in real time.[15] "If we had started a few years later, we would have missed it," she mused, and explained how the behavior would have been too widespread to offer the same research opportunity. (It's a lot harder to study how something is learned if everyone already knows how to do it.)

With the presence of culture and social learning in cockatoos now firmly established, Klump and her team are exploring additional questions. How exactly is the behavior transmitted from one individual to another? Why do male birds do the vast majority of the bin-opening? Does the behavior vary seasonally? Backyard observations continue to play a pivotal role, and have already revealed striking evidence of new innovations—not just in the birds, but also in the observers themselves. "People learn socially too," Klump reminded me, and described how neighbors had begun copying one another's schemes for protecting their bins from attack. Some tried to scare the cockatoos away with rubber snakes; others secured their lids with bungee cords, hooks, or heavy weights. And if the birds managed to overcome a strategy, people tried something new, with neighborhood patterns developing on both the human and bird sides of the equation. Klump called it an "innovation arms race," comparable—in a behavioral sense—to the interplay of adaptations and counteradaptations that often evolve between predators and their prey.

Barbara Klump began studying bird behavior more or less by chance, when a professor in graduate school happened to have an opening on a blackbird project. "Then I fell in love with birds!" she told me, and described how that first research experience blossomed into a career-defining fascination with avian learning, or, as she succinctly put it, "how problems are encountered and become solved." Klump's energy and obvious curiosity transcended our laggy Zoom connection, and she struck me as someone who probably spotted scientific opportunities—like parrots foraging in garbage cans—that other people might overlook. Case in point: carrion crows. When I asked about the backyard biology around her home in Germany, she told me that she'd noticed something intriguing about one of the most common birds in Europe. In short, she discovered the topic for her next major research effort by paying attention to nature on her daily walk to work.

"Every autumn, the crows are always dropping walnuts onto the road," she said, smiling and nodding, as if recalling a memory we both shared. In a sense that was true, since her story immediately brought to mind my frog-dropping robin. But where the robin went hungry, the crows fed well. Again and again, she watched them successfully use the pavement to crack tough nutshells and extract a tasty reward. Like the bin-opening cockatoos, Klump's crows had devised a clever strategy to access food that would otherwise be unavailable. The situation begged similar questions about how they learned to do it, and how those techniques might spread and diversify. Various nut-opening behaviors had already been observed in other crows, including a population in northern Japan known for placing nuts at intersections to be crushed by the tires of passing cars. Klump planned to investigate that larger cultural context, but she also saw nut-dropping as an opportunity to revisit conventional thinking on tool use.[16] "If a chimpanzee strikes a nut with a stone, we call that tool use," she said, citing an accepted norm in studies of animal behavior. "But if a crow drops a nut onto a stone, that's not tool use? It doesn't make sense!" Klump seemed perfectly comfortable with the idea that roadside observations of a common urban bird could challenge one of the major concepts in her field.[17] In fact, she sounded eager to get started. It seems that even trained professionals still find a lot to learn from backyard biology. And Barbara Klump is hardly alone in using backyard observations to study bird behavior. At the University of Gloucestershire in England, ecologist Anne Goodenough has applied strikingly similar methods to settle a long-standing debate about one of the most spectacular displays in nature.

"I've always been a bird-watcher," Goodenough told me, and explained that she became fascinated with murmurations the same way any bird-watcher would: by witnessing one. Named for the hushed rustling of their swooping turns and dives, murmurations are the enormous flocks of European starlings that mass on winter evenings, just before the birds settle into their night roosts. No one who has ever seen one will soon forget the hypnotic billowing of all those synchronized bodies, but scientists have never been able to agree on why it happens. Proponents of the "warmer together" hypothesis believed the habit to be a form of advertising, a prominent display meant to attract more birds and increase the number of warm bodies huddled at any particular roost. Others called it a "safer together" strategy, a way for the birds to reduce predation risk through sheer numbers, overwhelming the ability of raptors to pick out any one individual from the flock. People had studied murmurations for decades—counting the birds, timing them, filming them, and working out the physics of their synchronized flight. But the root cause of the behavior remained open to question, because no single research team had witnessed enough of the giant flocks to discern a pattern.

"The phenomenon happens quickly and is often not predictable," Goodenough wrote to me in an email. But it's also dramatic and eerily beautiful—hundreds or thousands of birds swooping and billowing in unison like animate smoke.

Whenever and wherever murmurations occur, they tend to attract attention, so Goodenough had no trouble finding willing participants for her online appeal, "#StarlingSurvey." In just two seasons, she amassed more than 3,200 detailed reports from across the United Kingdom and twenty-two additional countries. Starlings thrive in a wide range of habitats, so it wasn't surprising that people spotted the flocks everywhere from farm fields and woodlands to suburbs, city parks, and even industrial areas.[18] With such a bounty of data to analyze, Goodenough was able to quickly dismiss the "warmer together" idea. If the goal was a night roost with more birds and body heat, then murmurations should persist longer and attract more individuals on colder evenings when staying warm mattered most. But that wasn't happening. Instead, the larger, longer-lasting gatherings formed in the presence of predators, particularly when falcons, sparrow hawks, or other known bird-killers were actively harassing the flocks. Apparently, the starlings believed in a "safer together" approach. From a behavioral standpoint, their murmurations embodied the appeal of strength in numbers, which is not a bad way to describe Goodenough's approach to data collection. She continues to find surprising opportunities in the flood of everyday observations available through online platforms. Sometimes, it's not even necessary to write a survey or post a request for data. Mining old Twitter records for natural history comments has yielded highly accurate timelines for spectacles like the mating flights of ants or the sudden abundance of house spiders in autumn—whatever species and behaviors that people happened to be noticing and sharing with their friends and followers.

FIGURE 4.3. Thousands of neighborhood and backyard observations helped scientists show that murmurations of European starlings (*Sturnus vulgaris*) are more than beautiful—they help confuse attacking hawks and falcons, reducing the risk of predation. Image © Anne Goodenough.

While there is certainly power in numbers, our yards aren't just good places to crowdsource data collection. They are unique habitats in their own right, sometimes stimulating the very behaviors that experts find so interesting. As Klump pointed out, bin-opening cockatoos could never have developed without a city full of bins. Urban ecology is now considered a distinct field of study, focused on the many adaptations springing up in built environments that simply don't occur anywhere else. Not all are positive—invasive species, for example, often thrive in human-dominated landscapes, finding opportunities at the expense of native plants and animals. But a growing number of studies have documented local species embracing new habits, from bats and birds feasting on insects at streetlights to brushtail possums, stone martens, and chipmunks denning in artificial structures. There are even signs of potential evolution in action, including reductions in wing length for cliff swallows nesting under highway overpasses, apparently as an adaptation to improve maneuverability—survival—in traffic. Or consider the white-footed mice in New York City's Central Park, whose DNA is now measurably different from rural populations, showing changes associated with metabolizing a high-fat diet of junk food and peanuts. The number of research papers published annually with titles and abstracts featuring the word "urban" alongside "animals," "plants," or "biodiversity" has risen by more than 500 percent over the past two decades. With so many studies investigating the activities of wild creatures in heavily populated areas, the question isn't whether or not citizen scientists can contribute, but rather—as Klump and Goodenough both emphasized—how the work could possibly get done without them.[19]

In biology, unique behaviors can arise to suit any ecological setting. That makes the yards and byways of every neighborhood fertile ground for discovery, if only we remind ourselves to look closely, and, just as importantly, to look *everywhere*. Because even the most familiar surroundings contain unfamiliar perspectives. Chances are, no matter how hard we strive to be observant, there are fascinating habitats close to home that we've never laid eyes on before.

CHAPTER FIVE

Above

Send your ships into uncharted seas![1]

—Friedrich Nietzsche
The Joyful Wisdom (1882)

Six feet of taut elastic released with a snap like the crack of a rifle, launching our throw bag high into the forest canopy. I watched it disappear among the green boughs, a fist-sized, weighted pouch trailing bright orange line behind it like a party streamer.

"Looks like we might have got it in one," Tawm Perkowski said with satisfaction, peering up the sightline of his giant slingshot. He set the pole aside and used the throw line to haul a sturdy rope up and over the precise limb he'd been aiming for. Minutes later we were harnessed, helmeted, and ready to climb.

I was already glad that I'd hired a professional. My only previous tree-climbing experience (aside from occasional low forays in search of fruit) consisted of a one-credit course for arborists that I'd taken in graduate school.[2] I remembered little about it except for the final examination, where I thankfully hadn't even made it off the ground before the carefully knotted ropes and equipment I'd positioned in an oak tree came tumbling down at my feet. The instructor, looking on with a clipboard, didn't smile. In addition to a dismal grade, that embarrassment left me with the strong conviction that climbing tall trees should only be attempted in the company of an expert.

"I'll wait for you at the top," Perkowski said, and then hoisted himself quickly up the rope, squirrel-like, as if unencumbered by gravity. Moments later, I found the forces of physics fully and painfully in effect when I tried to follow him, lurching skyward inch by inch with a pair of mechanical ascenders. They're clever devices—sliding one-way along the rope and then grabbing tight for the lift. But the power is all manual, and I immediately understood why Perkowski, who might otherwise be described as "of average build," had the shoulders and torso of an Olympic wrestler. Still, the sweat and effort couldn't mask my growing excitement as I left the ground behind and entered a part of our yard that, in spite of more than two decades in residence, I had never explored.

The concept of forest canopies as a distinct habitat worthy of study didn't really take hold until the 1980s. One of the first clues came from below, when entomologist Terry Erwin spread tarps under nineteen trees in a "scrubby seasonal forest" in Panama and then released a choking fog of insecticide into the overhanging branches.[3] For beetles alone, the resulting rain of specimens produced roughly 1,200 different varieties, many of them found only in that particular species of tree. Erwin dubbed the canopy "the last biotic frontier,"[4] and predicted that thorough surveys would increase the world's estimated arthropod fauna from 1.5 million to more than 30 million species.[5] Epiphanies followed in other disciplines, and soon there were scientists of all kinds dreaming up new ways to access and study this unknown realm. Some climbed by rope, others erected steel observation towers, or built networks of aerial platforms, walkways, and zip lines. There were repurposed construction cranes transported piece by piece to remote locations, and even inflatable "canopy rafts" dropped into the treetops from hot-air balloons. Wherever and however they managed to ascend, biologists found that forest canopies were often strikingly different from the better-studied habitats beneath them—different in light and humidity and temperature, different in structure, and different in the community of species that called them home. Even familiar creatures often behaved differently in the trees, something I would observe almost as soon as I reached the top. But the first novelty I encountered as I heaved myself upward didn't grab my attention with flashy antics; it rubbed off on my shirt.

Unlike Perkowski's smooth ascent, my time on the rope involved a lot of awkward spinning, swinging, and bumping into the trunk of our target tree, a large Douglas fir. After one glancing blow, forty feet up, I noticed a luminous yellow powder speckling my arm from shoulder to elbow. Locating the source gave me just the excuse I needed to stop and rest, so I transferred my weight to the harness, grabbed hold of the tree, and found myself facing a brilliant variety of gold dust lichen that I'd never seen before.[6] It coated every furrow in a three-foot swathe of bark, so bright it looked unnatural, like a sprayed-on tag of fluorescent graffiti. Up close, I could see the individual grains that made up the crust, each speck containing enough of the species' characteristic mix of fungi and algae to start a new patch somewhere else. That's how lichens disperse from tree to tree: by tiny motes carried on the wind or the feet of birds or, for that matter, the sleeves of climbers. It's a chancy and haphazard process, so it takes a long time for species to accumulate. I shouldn't have been at all surprised to find an unfamiliar lichen growing on what was probably the oldest tree in our yard—any sapling might harbor a few common varieties, but only a mature trunk would have been around long enough to amass real diversity.[7] By some estimates, it takes more than five hundred years for fir trees in the Pacific Northwest to assemble the full suite of lichens and mosses capable of thriving in their canopies.[8] This tree was less than a third of that age, so anyone climbing it in, say, the twenty-fourth century will no doubt find a lot more things to bump into on the way up. They will also have a lot farther to go. I had chosen to climb this particular individual because I looked out at it every day from my office window, and had always admired how it soared above its neighbors, topping out at roughly 120 feet (36 meters). That was tall for our yard, but not by the standards of a Douglas fir. Given a long life and good growing conditions, the tree could eventually double—or possibly even triple—its current size.[9] If somebody does try to climb it in the centuries ahead, they will need to bring along

FIGURE 5.1. Some parts of any backyard are harder to observe than others. Here, the author hauls himself slowly upward into the crown of a large Douglas fir tree. Image © Eliza Habegger.

extra rope and a bigger slingshot, not to mention more upper body strength!

I apologized for taking so long when I finally caught up with Perkowski, but he assured me that he had been enjoying himself. "It's been years since I've climbed a tree and done nothing," he said. "I would normally have a chainsaw going by now!" Our climb was a far cry from the constant trimming, topping, clipping, and sawing of a typical day in his tree care business, but I told him that sitting still was precisely what I had in mind. To that end, he showed me techniques for harnessing myself safely to any branch that looked like a good perch, calmly explaining the ropes and equipment as if we were discussing them in a classroom instead of 95 feet (29 meters) up in a fir tree. I carefully repeated every procedure, determined to practice until my skills were, as Perkowski put it, a "muscle memory." But I couldn't help also noticing the scarlet dragonfly perched on a nearby twig, as well as another new lichen, this one coffee-brown and smooth, like an errant splash of latex paint. When at last I was strapped in and seated, and comfortable enough to really look around, I saw our yard spread map-like below my dangling feet: Raccoon Shack, garden, pasture, and house all suddenly revealed as part of a larger geography. There were glimpses of neighboring rooftops through the canopy, and broad views of a panoramic landscape invisible from the ground, where the forest stretched away to distant hillsides and patches of glinting water. It seemed hard to reconcile this sudden arboreal perspective with the yard I thought I knew—just what went on up here, unobserved, all day long? As if in answer, I suddenly became aware of the wasps.

Stinging insects come with the territory in the tree business. "I have run screaming like an eight-year-old through the woods," Perkowski told me later. And that was when he was lucky enough to be swarmed on the ground. Wasps also thrive in the canopy, where he once found himself trapped between two escape routes, forced to complete a series of complex rope maneuvers while under constant attack. "I finally made it down and took a look at my first eye sting!" he recalled, when I pressed him for the story. His use of the word *first* before "eye sting" did not escape me. But as I sat on that high bough, and saw just how many wasps were up there, weaving in and out amongst the green branch tips, a burning question overrode any worry in my mind. What on earth were they doing?

To clarify, the wasps in my fir tree hailed from a group broadly defined as yellowjackets. Most varieties belong in the genus known as *Vespula*, a taxonomic spin on *vespa*, the Latin word for wasp.[10] People in ancient Rome were very familiar with such striped, buzzing insects. Pliny the Elder's *Natural History* includes a whole chapter on wasps and hornets, subtitled with the telling description "Animals Which Appropriate What Belongs to Others." It's tempting to attribute that observation to the small thievery Pliny must have experienced dining alfresco, where—just as today—wasps would have been frequent visitors to the buffet. If so, then the old scholar certainly paid attention to what his harassers were making off with, noting "they are all of them carnivorous,"[11] or, in a more colorful translation, "all the sort of these live upon flesh."[12] Either way, that statement aligns closely with the modern understanding of yellowjacket biology. Living in highly social, queen-led colonies, they do indeed raise up large broods of carnivorous young. Most adults we encounter on the wing are workers, out scouring the landscape for insects and spiders to hunt, or scraps of meat that might be scavenged, all of it destined to feed the growing larvae back at the nest. Caterpillars rank among a yellowjacket's favorite prey, which is precisely what had me so puzzled. Because while I still had a lot to learn about the canopy of my Douglas fir tree, I knew for a fact that it was a crummy place to look for caterpillars.

More than 3,500 different moth and butterfly species make their home in the greater Pacific Northwest, flitting about in nearly every habitat, from coastal forests to riverbanks, prairies, mountain meadows, and more. Yet among all their varied caterpillars, only a handful can digest the tough, tannin-laden needles found on conifers. In one study, experts raised 158 regional butterfly varieties from egg to adult—studying every life stage along the way—and found a total of one species that fed on the foliage of Douglas fir. Those odds make a strong case that yellowjackets hunting in our yard would be better off at ground level, where butterfly caterpillars can be found munching away on everything from grasses (common ringlet, woodland skipper) to salal shrubs (brown elfin, spring azure) to stinging nettles (red admiral, painted lady, satyr anglewing, Milbert's tortoiseshell). Or if a wasp insisted on hunting in the canopy, then why not choose one of the nearby alder trees, whose broad green leaves host the caterpillars of at least six butterfly species and more than 200 different kinds of moths? No, the yellowjackets in my fir tree were clearly searching for something else, and if I wanted to know what it was I would have to start thinking, and seeing, as they did.

The compound eyes of a wasp include thousands of tiny hexagonal lenses, each one hardwired to the brain, where all those individual snapshots get assembled into a single, wide-angled view. Their vision is a kaleidoscopic blur that's hard for us to imagine, but we know that objects only come into clear focus up close, at a distance preordained by the fixed optics of those rigid, chitinous lenses. (Our own eyes, in contrast, have pliable tissues that constantly flex and adjust, providing visual clarity over a wide range of distances.) The yellowjackets foraging around my treetop perch hovered scant inches from the greenery, moving methodically among the branch tips. To see what they were seeing, I tried peering at twigs through a small magnifying glass. Like a wasp eye, it focused only at close range, forcing my attention toward the smallest details. I immediately encountered surprises: the spare beauty of the twigs themselves, velvety brown where new growth emerged from the husks of last spring's bud scales; the waxy smooth needles, sprouting at odd angles like stout grass, deep green above and lined below with stripes of pale stomata. What wasn't surprising was the lack of anything that seemed like a good meal for a wasp—no caterpillars, but also no flies, no sawflies, no beetle grubs. I did glimpse one tiny mite scurrying around, but that hardly seemed like something worth the effort of all that patient searching. It took careful examination of more than twenty branch tips before a new idea began to take shape. And then I found a telling clue.

At first I mistook it for a spider. Huddled in a defensive crouch, the leggy creature retreated backward from my lens, trying to hide itself in a dense patch of needles. Then I noticed its two eyes, reddish, and very unlike the eight black orbs that adorn the heads of spiders. The front "legs" were wrong too—not legs at all, but actually long, articulated antennae. When I spotted its piercing mouthparts, jutting downward like a needle-sharp straw, I knew that I was looking at an aphid—one of the giant conifer aphids, as I would later learn. I also knew that my growing suspicion was right on target. Prey might be scarce in the treetops, but all those green fir branches offered another resource with the potential to draw wasps from far and wide.

If Pliny the Elder's picnics were anything like the salmon barbecues of my youth, then they probably revealed another truth about the dietary preferences of wasps: in addition to a penchant for meat, yellowjackets have a sweet tooth. At barbecues, we learned to satisfy their carnivorous impulses by nailing the tail and backbone of the fish to a nearby tree. This was long before the invention of the plastic, scent-baited traps so common today, but it worked just as well—a tempting and undefended meal that helped lure the wasps away from the fishy portions on our plates. But that didn't mean it was safe to drink from a can of soda pop without double-checking the contents. More than once I heard an ominous buzzing as I raised a beverage to my lips, or watched wasps line up on the lid, sipping at the sugary drips pooled in the rim. The biology behind this habit is simple. Sugars provide adult wasps with the quick energy they need to fuel flight, foraging, nest maintenance, and other daily tasks—quite different from the protein-rich foods that larvae require for growth and development.[13] So when you see a wasp lapping up soda, it's feeding itself; a wasp scavenging salmon scraps is shopping for the kids. In more natural settings, hunting takes care of the protein side of that dichotomy, but where do yellowjackets get sugar? Some habitats offer flowers with abundant nectar, and wasps also target fruits and berries wherever they can find them. But there is another source of sweetness that is common in vegetation

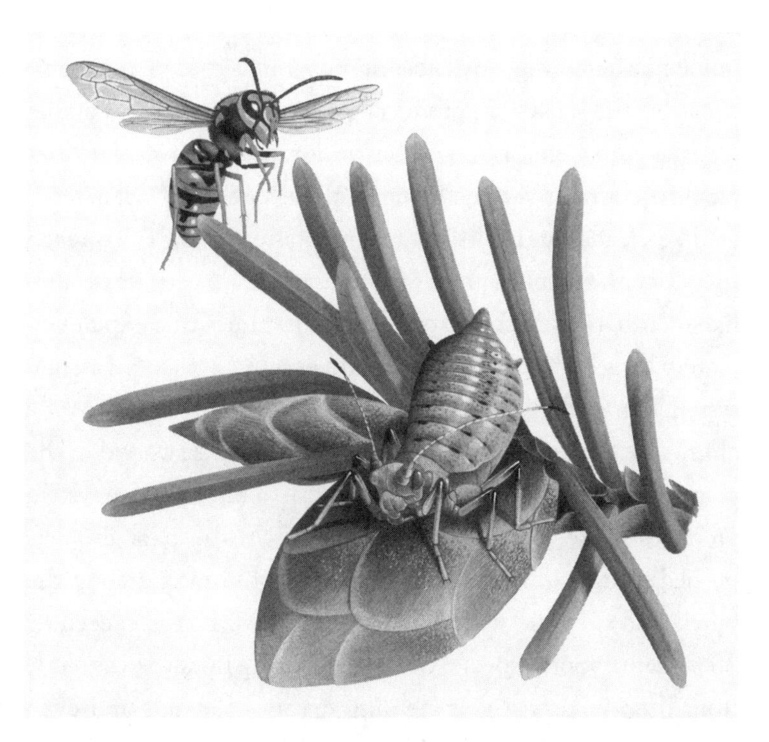

FIGURE 5.2. A yellowjacket wasp (*Vespula* sp.) approaches a giant conifer aphid (*Cinara* sp.) at the end of a Douglas fir twig, 95 feet above the author's backyard. Illustration © Chris Shields.

everywhere, even at the tops of fir trees, and few animals manufacture more of it than aphids.

Honeydew. The word really says it all—like dew, it collects on leaves, and, like honey, it is loaded with natural sugars. But where dew comes from moisture in the air *surrounding* plants, honeydew comes from the moisture *inside* them. It only ends up on the surface after passing through the gut of an aphid, mealybug, adelgid, or any of the other insects that make their living feeding on sweet plant sap. The trick lies in tapping into the phloem, a system of tiny tubes and channels that plants use to transport the sugar-rich products of photosynthesis. (A parallel system, the xylem, moves water from roots to leaves, but that pipeline contains very little sugar. Honeydew from xylem feeders like cicadas and spittlebugs rarely tastes sweet.[14]) To reach the phloem, an aphid probes at stems or leaves with its needle-like mouth as if searching for a vein. If its aim is true, sap immediately surges into the aphid's body because a plant's vascular system, particularly the phloem, is pressurized. This keeps the aphid's digestive tract filled for as long as the bite wound in the plant stays open, and, as everyone knows, a full digestive system produces waste. But, as waste products go, the clear droplets pushed from the nether regions of an aphid rank among the purest and sweetest, with sugar concentrations often exceeding 20 percent.[15] Some species of ants will lap up honeydew directly from the source, actively herding masses of aphids and even moving them from plant to plant, protecting their wards from attackers in a sort of mutually beneficial symbiosis. But most aphids in the world are untended, and their honeydew drips onto surrounding vegetation where its water content slowly evaporates, leaving a residue that is sweet, sticky, and free for the taking.

If my new theory was correct, then the yellowjackets in our yard—and presumably in any similar setting—were winging their way up into the canopy to feast on the honeydew of giant conifer aphids. This idea begged an obvious question. Were the twigs and needles on my fir tree noticeably sweet? A wasp could answer that query with taste receptors on its antennae and feet as well as its mouthparts, allowing a rapid assessment of any surface it landed upon. I only had my tongue to work with, but quickly learned that sampling vegetation near the tips of branches did indeed provide the occasional sugary reward. Admittedly, licking twigs in a fir tree sounds a bit eccentric, but in my defense, there must have been a time when people were far more familiar with the flavors of droplets on vegetation. Otherwise, who would have known to differentiate honeydew from any other kind of dew? Or consider the roots of the phrase "manna from heaven." Biblical scholars link that miraculous food, the salvation of the Israelites during Exodus, to the honeydew secreted by certain scale insects common on the Sinai Peninsula.[16] They infest tamarisk trees in such numbers that their honeydew often does rain down from above, hardened into pellets by the dry desert air. It is easily gathered, and one of at least half a dozen similar residues is still used in confections and traditional medicines throughout the Middle East. According to one study, honeydew products are referred to locally by the same term used for aphid or scale insect, a word that renders phonetically as "*mann*."

In a typical research project, this marks the point where I would plunge headlong into the current research, pore over a stack of relevant books and papers, and then seek out an expert to help explain exactly what was going on. Instead, a thorough search of the scientific literature taught me only one thing: Terry Erwin was right to describe forest canopies as a frontier. There were papers on everything from yellowjacket taxonomy to physiology, nesting habits, venom chemistry, and their sensitivity to various pesticides, but the habits of wasps in fir trees had apparently never been studied. And as for giant conifer aphids, the classic reference *Western Forest Insects* gave them only a single sentence, aptly summarizing the state of knowledge: "information on their distribution, abundance, and effects is scant."[17] I did find an intriguing series of papers from New Zealand, where invasive European yellowjackets have developed a voracious appetite for the honeydew produced by scale insects. They eat so much of it they've upended the ecology of local beech forests, displacing a range of native honeydew feeders, from other invertebrates to birds and even lizards. When I reached out to lead scientist Jacqueline Beggs at the University of Auckland, she returned my email within minutes. (Another hint that the world of wasp/honeydew research is rather small.) Beggs confirmed that the yellowjackets in New Zealand were active in beech tree canopies, particularly during the morning hours, when they would tank up on sugar before heading out to look for protein. She also said she was unaware of any published accounts of yellowjackets eating honeydew in western North America. "I'd be really interested to hear how you get on with your research," she wrote, and urged me to begin catching specimens in the treetops and examining their stomach contents. I might just do it—after all, if Tawm Perkowski can wield chainsaws in the canopy then I should be able to manage with an insect net.

Novelist and poet Hermann Hesse once wrote, "Trees are sanctuaries," an opinion he presumably formed with both feet firmly planted on the ground.[18] Climbing made that metaphor literal—whenever I ascended my fir tree, I was as dependent upon it for support as any of the lichens, beetles, birds, or other creatures inhabiting its branches. But Hesse's term is also a good fit in its classical Latin form, *sanctuārium*, the private royal chamber of a prince. Such places are rarely seen, so few people know their contents or the activities that take place within. What better way to describe the forest canopy, where even commonplace phenomena take on an aura of surprise? No one would call a rain shower unusual in the Pacific Northwest, but the first one I experienced in the treetops took my breath away. Instead of a distant patter overhead, the sound of the rain was completely immersive, a soft applause of droplets brushing branches and needles in all directions. It gave me a sudden appreciation of the canopy as three-dimensional space, a vertical topography of green that was by any measure the most expansive habitat in our yard. No wonder it still held so many secrets.

Given the challenge of safely ascending even a single individual, trees and their lofty reaches may seem destined to remain a "biotic frontier." But one doesn't have to climb them to appreciate them, or even to study them. Now that I'm paying attention, I hear the low hum of yellowjackets drifting down from the treetops all summer long, often beginning at the first light of dawn, just as Jacqueline Beggs described. Through binoculars I can spot their tiny yellow shapes, at least among the lower branches, allowing me to expand my observations far beyond a single fir. But ground-level learning about backyard trees can be even simpler.[19] Participants in the Treezilla project, for example, have mapped and measured more than a million oaks, maples, birches, and other urban specimens across the United Kingdom, often using little more than a smartphone and a ruler. Scientists can access those data to monitor things like tree cover, forest health, and the changes in spring flowering times driven by global heating. A similar program in North America, TreeSnap, amasses notes and digital photographs to track outbreaks of tree diseases and pests, and to help foresters locate and study healthy, resistant groves and individuals. Then there is the GLOBE Observer app, a NASA initiative that cleverly transforms smartphones into clinometers, allowing anyone to calculate canopy height at any location. Why? As a crucial safeguard for ground truthing and correcting land cover data beamed down from satellites. But by far the most popular citizen science tree platform in the world, available everywhere and cited in thousands of articles, government reports, and peer-reviewed publications, addresses a different sort of question altogether: How much are backyard trees worth?

"Dollars are the great equalizer," David Nowak told me over the phone, and then explained that putting a monetary value on trees had never been his original intent. "I wanted to know what trees do," he said, describing his initial curiosity about urban forests as purely scientific. "We knew we had X number of trees in city Y," he went on, recalling the research landscape early in his career, "but what does that mean in terms of biological, chemical, and physical processes? It really hadn't been studied." As a senior scientist for the US Forest Service, Nowak was in a position to find out. He began projects on how trees in cities impacted things like air pollution, carbon dioxide levels, ozone levels, and more, but quickly encountered what he calls his greatest pet peeve. "We have all these scientists doing science," he said, "but no one is *using* the science." That disconnect between knowledge and action led directly to the development of i-Tree, a user-friendly software package for translating complex tree biology into something that anyone can relate to.

"Keep the measurements simple and let the models do the work," Nowak said, summarizing the philosophy that has guided i-Tree's development and expansion for more than fifteen years. The result is a tool that needs only the address, species, and circumference of a tree to calculate its impact on things like air quality, temperature, or—a popular option these days—carbon sequestration. "We can value that," Nowak assured me, explaining how i-Tree accessed vast troves of Forest Service monitoring data to generate precise growth estimates for any tree in any location. And if you know a tree's circumference and how fast it's growing, you know how much carbon it is stocking away in its wood, not to mention how much shade it provides, and how many leaves it has to cycle air, retain water, muffle noise, and provide all the other benefits of vegetation in built environments. "I like to call it the green envelope," he said. "When you walk outside, how much are you enveloped by green?"

In conversation, David Nowak talks fast, as if his mind has learned to keep pace with the i-Tree algorithms he helped to develop. Now retired, he still works on the project as a volunteer, eager to continue refining and expanding the i-Tree tool kit. "How do you value people's emotional responses to trees?" he mused. "How do you put a value on aesthetics, or biodiversity?" The challenge of capturing such intangible assets means that i-Tree's calculations will always be on the conservative side. But the numbers are still impressive enough to attract attention.

"We have over 600,000 users in more than 130 countries," Nowak told me, and then he became even more animated describing how people were putting i-Tree to practical use, from reforestation schemes to the calculation of carbon credits to neighborhood tree-preservation efforts. "People have gone out and hung big price tags on the trees!" he enthused. But i-Tree's software can do a lot more than generate price tags. The ultimate goal is what Nowak described as "an intelligent management system." Features in the works will allow site-specific risk assessments for threats from fire, drought, pests, and disease. There will be tools to identify the best species to plant at any location, not only now but also in the future as the climate warms. "It will tell you what you have, what you need, *and* how to sustain it," Nowak said, and then emphasized his hope that i-Tree could inspire something more fundamental: curiosity. He envisioned people taking a new interest in trees—how to manage them, but also how to better appreciate everything they do for us. In other words, putting dollar values on trees is just a starting point, or, in Nowak's words, "a gateway drug."

The day after I spoke with David Nowak, in what can only be described as a case of storytelling serendipity, an inspector from the local county assessor's office stopped by unannounced to update the official appraisal of our property. He asked about a shed under construction and the new roof on the barn. We discussed the rising cost of septic systems, driveway gravel, and other material improvements that might impact the property's assessment for tax purposes. The value of trees never came up. Later, I took a flexible measuring tape into the woods to put a few numbers on what the county was overlooking. The circumference of my big climbing fir came in at 99 inches (251 centimeters), giving it an i-Tree value of $173.45. A smaller pine nearby was worth $59.38, and the alders at the edge of the meadow averaged $24.03 each. With several thousand other trees crowding our woods, and more coming up wherever we stopped mowing the grass, it didn't take an algorithm to do the math. Forget the shed and the driveway, and even the house; the trees were clearly the most valuable asset on our land.

Property owners can always challenge an appraisal they disagree with, and i-Tree's calculations are similarly open to question—they depend on assumptions about things like carbon sequestration, pollution mitigation, and the retention of rainwater during storms. But debating such details is part of the point. By simply taking the time to assign specific values to specific trees, I had begun thinking differently about the firs, pines, and alders in our yard, and that's precisely what Nowak and his colleagues at i-Tree had in mind. Between that and my climbing forays into the canopy, it's safe to say that the arboreal aspects of our property had begun to seem a little less remote. And that got me thinking about another ubiquitous backyard habitat that is even less well-known than trees. Fortunately, the only tool required to explore it can be found in virtually any garage or garden shed.

CHAPTER SIX

Below

No, up and out of doors is good enough to roam about and get one's living in; but underground to come back to at last—that's my idea of home!

—Kenneth Grahame
The Wind in the Willows (1908)

Facilitators at conferences or group retreats sometimes pose "desert island" scenarios to help break the ice. If you could only have three foods on a desert island, for example, what would they be? Or what books would you choose? What music? Endless variations make it a nice way to start conversations among strangers. Biologists play a similar game with a question about equipment: If you could take just one tool into the field to do your research, what would it be? Binoculars probably rank as the most common choice, although some people prefer a hand lens or even a portable microscope. Over the years, I've listened to cogent arguments for a thermometer, a measuring tape, and a good digital camera, but there is one answer that forever changed how I think about landscapes. I heard it only once, from a bearded naturalist on the forested shoreline of a pond in Vermont.

"A shovel," Brett Engstrom said without hesitation. He was leaning on one at the time, which added a certain authenticity to his response. Rangy, weathered, and totally at ease in the woods, Engstrom held an esteemed status in New England's botanical circles, famed for his uncanny ability to locate and describe rare plant communities. As a student at a nearby university, I was thrilled when he agreed to lead a field seminar for the members of my graduate program, and expected to spend the day identifying species, and learning which ones tended to grow together. We did those things, but only after Engstrom used his favorite tool to dig a hole at every site we visited and examine the soil. He urged us all to do the same thing. Nothing else integrated so much information about a habitat, he told us, from geology to climate, hydrology, biological activity, and more. Learn the soils, he explained, and they will tell you what plants to look for.

Decades later, I would like to say that I'd taken Brett Engstrom's advice. But while I remember that huckleberry shrubs and rhododendrons like acidic soils, and that ash trees thrive in rich, sandy loam, I can't say that a shovel has ever featured prominently in my cache of field equipment. At home, the only time I remember digging a hole to examine the soil was when we were choosing a place to put the garden. That decision, in various forms and with various tools, has been playing out ever since the dawn of agriculture. Farmers everywhere know the look and feel of the different soils on their land, and have learned by long experience which ones will best support a particular crop. Published soil classification systems date back at least 2,500 years, to the *Yu Gong* in the Chinese *Book of Documents*, which recognized nine major soil varieties and established a land taxation scheme based on levels of fertility.[1] Given such ancient ties, it's surprising that it took so long for the underground realm to be recognized not just as a place for roots to take hold, but as a distinct and important habitat, filled with its own suite of unique species.

Until well into the nineteenth century, scientists regarded soil as little more than a medium, a temporary storage place for the nutrients used by plants. Scottish geologist James Hutton neatly captured the prevailing attitude in his 1788 treatise, *Theory of the Earth*: "a soil is nothing but the materials collected from the destruction of the solid land."[2] Once again, Charles Darwin played a role in changing conventional thinking, this time with his backyard studies of earthworms. Between 1837 and 1881, he published a series of articles and a popular book on how worms aided in the weathering and mixing of topsoil, or, as it was then known, "vegetable mould."[3] Russian scientist Vasily Dokuchaev went on to formalize the idea of soil biology as a distinct field of study, citing Darwin's work but going further to argue that the entire community of subterranean organisms at any given location contributed to the formation of its distinct soil type. Both scholars were on the right track, but it would take advanced microscopy, genetic fingerprinting, and other modern tools to truly unlock the richness and complexity of life underground.[4]

The numbers are hard to fathom. Bacterial diversity in a single gram of soil has been measured at more than fifty thousand distinct species. Expand that sample to a spadeful and you can add in two hundred different kinds of fungi, dozens of worms and nematodes, at least a hundred kinds of mites, twenty varieties of springtails, and over a thousand different amoebas, flagellates, and other protozoans. And those are just the creatures that live there full-time. Many others pass through for at least a part of their life cycles, from slugs and fungus gnats to various bees, beetles, and burrowing mammals. Add to that an untold host of dormant seeds, and some truly strange things like tardigrades and pseudoscorpions, and it's clear that any patch of ground holds as much life below the surface as above it—often more. The vast majority of that diversity has yet to be described or studied, making every loamy gram its own potential biotic frontier. But unlike the forest canopy, which can seem like a world apart, the soil is constantly underfoot, filled with hidden activities that influence life aboveground in ways scientists are just beginning to understand.

With that in mind, I found myself trying to see our yard as someone like Brett Engstrom might, or, for that matter, Darwin or Dokuchaev. Would conditions belowground explain why twinflowers grew in only one place in our woods, clustered around the base of an otherwise nondescript young fir tree? Did soils control the distinct line where our conifers transitioned to a forest of alders? Had they informed an earlier generation, telling farmers where to clear out the trees entirely for pastures and crops? And what sorts of creatures were living down there, out of sight beneath our feet all day long? On a sunny morning in early autumn, I borrowed a shovel from the garden shed and headed into the woods to see what I could learn.

Just off the path and past a pair of old fir stumps, I found a mossy, flattish place and set to work. I had chosen to dig in the forested part of our yard because its soils seemed the most pristine—never plowed up for crops, or excavated to build a pond, or shoved around and flattened to make a site for a home. The ground remained hummocky and rough, crisscrossed here and there with fallen logs and branches in varying states of decay. When the first plunge of my shovel sparked off a large rock, I began to understand why this part of the property had never been farmed. More rocks followed, but in time I began making steady progress, scraping out spadefuls and laying the spoils carefully on a tarp beside me.

Professional soil scientists dig tidy, sharp-edged pits deep enough to stand in, and competitors at the US National Collegiate Soils Contest square off in trenches neatly carved by an excavator. The hole I ended up with looked more like the work of a drunken badger, but it was deep enough to sample four distinct types of material—several inches of duff on the surface, a chocolaty brown layer beneath, something rockier and colored like coffee and cream below that, and a pale mixture of stones and sand at the bottom. Judges at a soil contest would call these layers "horizons," and point out that the chocolate color came mostly from decomposed plant matter working its way downward, while the coffee area held more minerals and clay, some of it weathered in place and some accumulated from above. (Most soils also have a sandy "eluviation horizon" in between the chocolate and the coffee, but it's often thin, and I couldn't spot it.) The deepest part of my hole hit on "parent

FIGURE 6.1. Four distinct soil horizons are visible in this stony loam, dug from the wooded portion of the author's yard. Less visible are the myriad fungi, bacteria, protists, nematodes, and other tiny organisms that make soil among the richest biomes on the planet. Image © Thor Hanson.

material," the original substrate, in this case a mix of rock, sand, and silt left behind by a retreating glacier some fourteen thousand years in the past. It was that long-ago moment, when the last ice melted away and exposed barren, rocky ground to the elements, that soil formation began in our yard.

While we often think of soils "building up" over time, much of the activity actually goes in the other direction. Precipitation percolating downward helps weather the parent material, freeing up minerals and acting as a conduit for the dead organic matter that begins to accumulate as soon as lichen, fungi, and simple plants become established. That water also helps form a matrix for the bacteria and myriad other tiny organisms that arrive and begin contributing to decomposition and nutrient cycling. By the time larger, rooted plants take hold, the soil hosts a complex community in an increasingly complex environment, where water, air, minerals, and living things interact to anchor everything we take for granted at the surface. Much depends upon the type of parent material, the topography, rainfall, and other local factors, leading to enormous differences in depth, texture, color, fertility, and other key traits, even over short distances. Within a few hundred yards of my forest pit, for example, I later found extremely sandy parent material below our garden, explaining why its topsoil is comparatively deep and well-drained. Near the pond, I encountered a horizon rich in clay that apparently trapped water close to the surface, favoring alders and willows over fir trees. Such variations are site-specific, but the processes of soil formation are universal, which means that champions from national competitions around the world find themselves on equal footing at the International

Soil Judging Contest, an Olympic-style event held every four years in locations (and biomes) as far-flung as Scotland, Brazil, South Korea, and Hungary. If the competition were held on our property, participants would probably agree with the government soil scientists who passed through in the 1950s, mapping our land as part of a countywide survey effort. "Stony loam," they concluded for the soil type I'd been digging in, echoing the management decisions of generations of farmers when they described it as ill-suited for crops, but promising for the growth of Douglas fir.[5]

Academically, I understood the basics of soil biology, and the pivotal role that soils play in terrestrial ecosystems; but the fact remained that, even when arranged artfully on a tarp, the stony loam from under our woods looked more or less inert. It was easy to see why people have so often dismissed soil as little more than fractured earth. What was it that made such an uninspiring habitat home to so many species? To answer that question, I turned to the only person I knew of who had ever tried to count them.

"This took a long time!" Mark Anthony told me in an email, describing months of effort ferreting out data from published and unpublished research projects around the world. As a fungal ecologist at the Swiss Federal Institute for Forest, Snow and Landscape Research, Anthony had expected a lot of information on mushrooms and molds. After all, fully 90 percent of the world's fungi are soil dwellers. But it turned out that most of the world's bacteria also live in soils, along with the majority of plants and large portions of such speciose groups as mites, nematodes, ants, millipedes, and protists. When he and two colleagues added up the numbers, they found that soils give home to a staggering 59 percent of *all* species at one or more stages of their lives, making the humble ground beneath our feet far and away the richest biome on the planet.

"Soil supports so much biological diversity because it is also very physically and chemically diverse," Anthony explained. To small organisms, subtle variations in traits like particle size, moisture, or acidity represent radically different environments, each one with its own evolutionary opportunities for adaptation and specialization. Many soil creatures are also what Anthony called "dispersal limited." They don't move around much, a quality that, combined with their minute proportions, lends itself to high diversity in tight quarters. "Even organisms with similar niches can often co-exist," he wrote, "because they may never encounter each other."

Paradoxically, the same issues of scale that allow so many life-forms to thrive in soil also make those life-forms incredibly difficult to appreciate. "I think about it a lot," Anthony responded, when I asked how a layperson could best learn to recognize and value the soil biodiversity in their backyard. He directed me to several websites that featured vivid, highly magnified images of various soil dwellers, including glassy, translucent potworms and a shiny orange mite with bizarre front legs four times the length of its body. But he also acknowledged that "visualizing these organisms is really not enough on its own because most soil life is way too cryptic to be imaged or perhaps not conventionally attractive enough to people." Better storytelling may provide part of the solution, he suggested, pointing out that people have learned to care about all sorts of creatures they'll probably never lay eyes on, from polar bears to blue whales to orangutans. But the best way to start appreciating soil biodiversity probably lies with the species that *can* be seen, and that requires a bit of hands-on exploration.[6] "People can totally become empathetic for soil life," Anthony argued, just as soon as we all spend more time poking around in the dirt.

With Anthony's words in mind, I lay down beside my soil pit and began sorting through the top layers of material. Up close, the ground smelled suitably earthy, like root vegetables fresh from the garden, and I could see how the feather mosses formed a mat interlaced with dead leaves, fir needles, and lichen-crusted twigs. Peeling back the greenery, I immediately spotted a shiny brown spider, scuttling for cover, and then something small, dark, and curled that I took to be a springtail when it suddenly uncoiled itself in an explosive leap, vanishing from sight. There were more spiders as I probed the zone where rotting vegetation mixed with soil, and then I uncovered a bronze-colored burrowing bug, plodding along like an insect rhinoceros. Through my hand lens, its compact, armored body looked pockmarked with round perforations, and I wondered if they served any purpose. Bark beetles use similar dimples to transport fungal spores from tree to tree, spreading the very agents of rot that soften the wood they feed upon. That seemed unlikely for burrowing bugs, which are mostly sapsuckers that tap into living roots for sustenance. But something must have been moving spores around, because no form of living thing was more ubiquitous in those top inches of soil than the pale, twining hyphae of fungi.

FIGURE 6.2. Advances in macrophotography have enabled us to put a face on many forms of seldom-seen soil life, including podgy springtails (*Holacanthella duospinosa*) (top), fluid-feeding mites (*Linopodes* sp.) (middle), and pincushion millipedes (*Polyxenida* sp.) (bottom). Images © Frank Ashwood.

In college, one of my favorite professors kicked off his evolution course by showing the class a vintage illustration of the Tree of Life, the sort that featured "Man" at the pinnacle of the lofty crown, with various "lesser" life-forms scattered around on the lower branches and twigs. He gave us a few minutes to study the image and then started pointing out various taxonomic groups and lineages, contrasting their dated arrangement with the modern perspective. Tellingly, fungi had been left off the tree altogether. Just as tellingly, not a single student noticed. "It's only a kingdom!" he thundered, berating us for overlooking a category of life as fundamental as plants or animals. Later, he told me that it was the same every semester: even in a room full of aspiring biologists, few people gave more than a passing thought to fungi. Fortunately, a lot has changed in the years since. With better detection tools (and a better understanding of where to look), mycologists have been discovering new fungal species at a dizzying rate, and not just in soils. Diverse and prolific fungal communities thrive everywhere from seawater to solid rock to the insides of plants and animals (including us), playing vital and often symbiotic roles in nutrient cycling, decomposition, and more. And as our scientific understanding of fungi has grown, so too has their public profile, from bestselling books and award-winning documentaries to an ever-expanding range of products and potential applications. Mushroom leather, anyone? Mushroom coffee? How about fungal-based biofuels, fertilizers, meat substitutes, building insulation, plastic recyclers, or self-healing, fungal-infused concrete? By any measure, it's fair to say that fungi are having a moment. It's also fair to wonder, seeing how ubiquitous their mycelia are in the average soil pit, what took so long?

"Fungi everywhere," I jotted in my field notes, and then added an important detail: "Seemingly twined with roots." Here, then, was visible evidence of perhaps the most famous of all recent fungal revelations: how they form a symbiotic network of hyphae and roots, connecting many (if not all) the trees in forested habitats. Popularized as the Wood Wide Web, this underground mat serves as a great marketplace, with fungi and trees exchanging nutrients, water, and photosynthetic sugars—a relationship still filled with mystery.[7] Ask a botanist, and they might tell you that the trees use the fungi to share resources among neighbors and closely related individuals. Ask a mycologist, and they might tell you that the fungi are the ones in charge, strategically shifting resources among the many trees in their thrall. Ask a fan of science fiction, and they might call the situation a good premise for a zombie apocalypse story—just replace the trees with people, and you have the popular television series and video game franchise "The Last of Us." What is beyond dispute is how this discovery is changing the field of forest ecology, replacing old concepts of strict competition among trees with the possibility of cooperation, communication, and, in the most daring interpretations, compassion. Research continues in laboratories, field stations, and deep woods, but the basics can be seen by anyone with a shovel and a wooded yard. In fact, any vegetated backyard habitat may suffice. Evidence is mounting for the existence of similar networks among soil fungi and plants in all sorts of terrestrial environments, from savannas and prairies to arid shrublands, and even arctic tundra.

Before filling in my soil pit, I took samples from each horizon and layered them up in a small, clear vial, like a tiny earthen parfait. I keep it now on the windowsill in front of my desk. Framed against the view of the yard, it serves as a constant reminder of the unseen world below. I also filled up a specimen bag from the upper layers and shipped it off to a laboratory at the University of Oklahoma, on the not-inconceivable chance that it might contain a cure for cancer.

"Hang on, let me grab something," Robert Cichewicz said, and disappeared from the screen of our Zoom call. I could hear him rummaging around somewhere out of sight in his office, a bright and airy room in the Stephenson Life Sciences Research Center at the University of Oklahoma. Then he was back, holding a long scroll of paper filled with maps. "This is where we started," he said, showing me an outline of the United States sprinkled with a few dozen dots, many of them clustered around Oklahoma City. "At first it was mostly friends and family we cajoled into doing it," he went on, explaining how he and several colleagues had sent out an online appeal asking people to mail in soil samples from their backyards. "We had this idea of doing natural products discovery from soil fungi," he said, but none of them had the time or resources to travel the country, filling little baggies with dirt. A few people found their website and pitched in, but the project didn't really take off until it got noticed by a popular influencer on Reddit. "Then we had our viral moment," Cichewicz recalled, sounding as if that fact still surprised him. The next map he showed me was crowded with dots from coast to coast. "We've had over twelve thousand participants!" he crowed.

Of course, receiving all those samples was just the beginning. Each one had to be individually processed, a lengthy task that involved isolating unique fungal strains from colorful colonies grown on petri dishes, and then culturing the isolates on their own beds of something that looked strangely familiar. "They're Cheerios," Cichewicz confirmed, rattling a test tube filled with breakfast cereal in front of the camera. Apparently, the same toasted oat rings that feed so many North American people in the morning also make a good meal for North American fungi. After three weeks on a Cheerio diet, the isolates get whizzed up and soaked in a way that reveals all the chemicals they've been producing in their digestive and other metabolic activities. "It's like making tea," Cichewicz explained, "but with solvents instead of water." Considering that the average soil sample yielded five or ten culturable fungi, and some had closer to fifty, the numbers quickly exceeded normal working capacity for an average university biochemistry lab. Keeping up with the continuing flood of contributions suddenly required a whole new level of staffing and funding. Fortunately, Cichewicz and his team found eager collaborators at a place where that scale of project was routine: the National Cancer Institute (NCI), one of the largest scientific research agencies of the US government. Also, fortunately, despite their large budget and thousands of employees, not everything at NCI was state of the art.

"Their microbial collection was crap," Cichewicz observed cheerfully, saying it with such disarming candor that I could easily imagine him telling the officials at NCI much the same thing. However that conversation played out, Cichewicz and his team ended up with a contract to rebuild the fungal component of the NCI research collection from the ground up (in every sense). "It brought in revenue for the project," Cichewicz allowed, but he seemed more excited about how the contract had made tens of thousands of backyard fungi, and all their associated chemical extracts, "permanently available to a wide array of researchers." Yes, searching for new cancer drugs was a priority, but anyone with a viable idea could access the collection, from those investigating other diseases like malaria and multiple sclerosis to scientists working on soil fertility, bioremediation, natural pesticides, and more. "It's a perfect match," Cichewicz summarized. "They're biologists. They need compounds to test. We're chemists. We have compounds!"

Lanky and bespectacled, with a neatly trimmed, gray goatee, Robert Cichewicz fits the image of a distinguished, midcareer professor. But when he talks about soil fungi he seems on the verge of shouting with excitement, like someone dying to let you in on a big secret. (Among all the experts I spoke with for this book, Cichewicz was the only one who didn't just say yes to an interview, but went so far as to set up the Zoom call himself and send me a link.) "Fungi are huge ecological drivers," he enthused, citing their vital roles in rotting and recycling all manner of detritus, "but there are a huge number of unknown organisms out there, and even among the known species, we really don't know their chemistry." The fact that backyard samples had a role to play in changing that equation pleased him scientifically, but he also saw it as a way to spread his passion and curiosity. Participants in the project could access their samples online and download high-resolution images of their own yard's fungi growing on petri dishes,

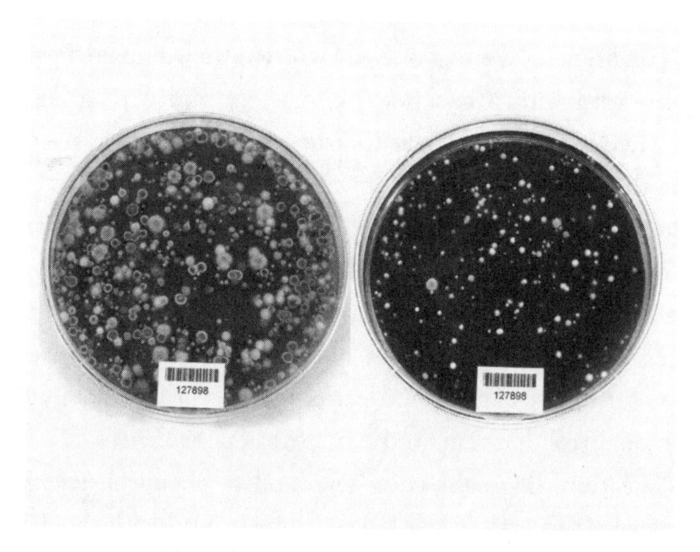

FIGURE 6.3. The soil sample from our yard contained sixteen distinct fungal strains, shown here growing on petri dishes in Robert Cichewicz's lab at the University of Oklahoma. Image courtesy of University of Oklahoma Citizen Science Soil Collection Program.

something I had done moments before our conversation. "Not bad!" he congratulated me, when I told him that my sample had produced an above-average total of sixteen distinct fungal strains. I could expect to hear more from him later if anyone made a breakthrough studying my contributions to the collection.

Cichewicz said he could think of eighty or a hundred scientists currently working on fungal compounds derived from backyard samples, with dozens of promising discoveries already in the books. And whenever a new paper is published, he and his colleagues make sure the original soil contributor is acknowledged, and they get in touch with that person to share the good news. "We had one guy worried we would be digging up his yard with a backhoe!" Cichewicz said with a laugh. But once he makes it clear that no additional soil is needed from their yards, people are typically thrilled to have provided something useful.[8] Perhaps no one has more claim to bragging rights than the woman from Salcha, Alaska, a small town near Fairbanks, whose baggie of fungi-laden soil has already produced two major discoveries: maximiscin, a potential treatment for breast cancer, and pericosine, a curious metabolite with qualities that remain the subject of debate.

"We think it might be a new kind of chemical defense," Cichewicz speculated, and held his hands out like little claws, describing a molecule that seemed to lie in wait, ready to attach itself to any incoming threat. Biologically, it's a reminder that competition, predation, and other adversarial interactions in soil often play out chemically, which is partly why fungi produce so many intriguing compounds in the first place. They're involved in a constant evolutionary arms race with their neighbors, fungal and otherwise, trying to carve out and hold territory by developing novel means of attack and defense. What makes pericosine so unusual is its ability to neutralize almost anything that other fungi, or curious scientists, have tried to throw at it. Scaled up, Cichewicz predicted that this knack might have important uses in the human world, such as quickly disabling and detoxifying chemical weapons on the battlefield. But his university superiors balked at the idea of getting involved in military research, so he decided he needed to find a less controversial study subject.

"What about skunks?" he reasoned. "They can't be against skunks!" And so, with no objections from the higher-ups, Cichewicz got his hands on some skunk scent and set to work. "Have you ever smelled a skunk gland?" he asked, wrinkling his face. "It's the nasal equivalent of looking at the sun." But when he and his team added a drop of pericosine, the stench disappeared in less than a second. Tests with other putrid animal products met with similar results. ("Who knew you could buy wolf urine on the internet?" he mused.) The project might have stopped there, but a local company got wind of it and saw the potential for using pericosine on the smells associated with a different set of animals. With major retailers already showing interest, an all-purpose pet-deodorizer may be the first commercial product brought to market from Cichewicz's vast library of samples.

New pharmaceuticals face higher regulatory hurdles, but dozens of backyard fungal compounds (in addition to maximiscin) are now under study. If any of them bear fruit, it wouldn't be the first time that a medical breakthrough came from a bag of dirt. In 1943, Rutgers University student Albert Schatz isolated the lifesaving antibiotic streptomycin from a microbe he found in some "heavily manured field soil" near his lab at the New Jersey Agricultural Experiment Station.[9] The discovery led to the first effective treatment for tuberculosis, and helped earn Schatz's thesis advisor, Selman Waksman, the Nobel Prize.[10] Waksman's lab went on to develop other soil-derived antibiotics, starting a trend of discovery that has never abated: more than two-thirds of modern antibiotics still trace their origins to bacteria and fungi living in soil.

Building on the success of the University of Oklahoma project, several other research groups have also begun crowdsourcing backyard soil samples, including the Drugs from Dirt program at Rockefeller University in New York and Antibiotics Unearthed, a project run by the Microbiology Society in London. But Robert Cichewicz hopes for a lot more than basic citizen science. He wants participants to keep thinking about soils long after they've sent in their samples and seen a picture of their fungi growing on a petri dish. He wants the experience to spark a new way of thinking about the ground we look at and walk across day after day.

"Don't treat your lawn like a countertop," he told me, when I asked what people should do to better appreciate their soils. "It's not to be cleaned and dusted! Let it go wild as much as you can tolerate, and then watch what is going on." Growth rates, patches with different colors, the establishment of new species—all such patterns are probably driven or influenced by fungal activity belowground. "If you see a mushroom, go take a gander. Take a picture. Enjoy it for a moment," Cichewicz encouraged. "Because I guarantee you, there's a lot more going on underneath."

Go take a gander. That seemed like good advice for learning about any backyard habitat. I kept it top of mind when I finally turned my attention to another part of our property that I had been looking at daily for years, apparently without seeing a thing.

CHAPTER SEVEN

All Wet

And then, all of a sudden I had the revelation of the enchantments of my pond.[1]

—Claude Monet
À Giverny, chez Claude Monet (1924)

The human brain is a master of improvisation, creating our waking reality through an unconscious combination of fact and fabrication. What we see, for example, does not truly reflect the flood of visual data entering our eyes at any given moment. That much information would overwhelm the senses; it simply can't be processed fast enough. Instead, the brain plucks out a few real-time details from the center of our attention—faces and shapes, perhaps, or the speed of an object in motion. The rest of the picture gets cobbled together in an instant, drawn from impressions and previous experience, as if the brain were a jazz soloist perpetually inventing variations on the fly. That helps explain why witnesses to a crime can't always agree on the color of the getaway car, or why moviegoers rarely notice the body doubles that stand in for cast members during action scenes. And it's almost certainly why it took me more than ten minutes to realize that I was staring at a snipe.

The window above our kitchen sink overlooks the deck, the corner of the house, and a patch of mowed grass along the edge of our small dug pond. Willows and cattails frame the water, and there is a row of tall fir trees in the distance that catches evening sunlight long after the rest of the yard is in shadow. Hand-washing the dishes gives us ample time to enjoy this view. It's the sort of scenic backdrop to daily life that I'm sure my brain usually fills in by rote. So perhaps it's not surprising that I had nearly finished cleaning up after breakfast one morning before a brown smudge at the water's edge finally coalesced into the shape of a bird. It had been standing there in plain sight the whole time, stock-still, a plump brown form with a comical bill that stretched out half again as long as its body, like something drawn on by a child. In my brain's defense, the Wilson's snipe is notoriously difficult to see, so cryptic that many people have come to think of it as a myth.[2] The phrase "snipe hunt" in North American slang refers to a folly with even less chance of success than a "wild goose chase." But it was precisely the bird's secretive nature that made this such a rare opportunity. I dried my hands, grabbed a pair of binoculars, and slipped quietly out

FIGURE 7.1. The Wilson's snipe (*Gallinago delicata*) haunts wet meadows and thickets from Venezuela and Colombia north to the arctic, sometimes visiting backyard ponds during migration. Seldom seen, it may be best known for the eerie winnowing sound made by its tail feathers during mating flights. John James Audubon (c. 1827), Rawpixel.

onto the deck to watch. It quickly became apparent that I had been missing a lot more than a snipe.

From my perch on our wide deck railing, I had a clear line of sight to the bird. But as soon as I focused my binoculars, I realized that something was amiss. The plumage looked too dark for a snipe; the eye was reddish; and the bill—though still comically large—wasn't nearly long enough. Somehow, the snipe had transformed itself into a Virginia rail! My brain struggled with this development until the rail moved suddenly to the left and I spotted the snipe behind it, still standing like a statue at the water's edge. Then *another* rail appeared, dashing out from the reeds and into the shallows where it astonished me by alighting on the surface of the water and swimming in a tight circle before returning to shore. The first rail was still darting back and forth in the short grass, apparently snatching insects from the ground, while the snipe had begun taking slow, deliberate steps, bent over like a patient beachcomber, probing at the shoreline mud with its bill. And just like that, in a matter of seconds, I had witnessed more snipe and rail behavior than I would usually observe in a year, prompting the rare use of an exclamation point in my field notes: "Three impossible-to-see birds all together in full view!"

Biologically, the elusiveness of snipes and rails is a straightforward matter of habitat. Both species spend much of their time in "rank vegetation," the combination of tall grasses, sedges, rushes, and shrubs that so often dominates in wet places, from ditches and low-lying fields to the edges of ponds, marshes, and streams. Rails are particularly specialized for this lifestyle, with durable head feathers adapted for pushing through dense vegetation, and narrow, "laterally compressed" bodies that hardly make a rustle as they pass. (The familiar phrase "skinny as a rail" refers to this bird's distinctly flattened appearance, strong evidence that the features of backyard birds—even the secretive ones—were once common knowledge.) Primarily ground dwellers when not in migration, both species use shyness and camouflage as a form of defense; better to stay hidden in the weeds than risk exposing themselves to hungry predators. That threat came into sharp focus moments later, when a red fox suddenly appeared, loping across the lawn, directly toward the birds.

One rail immediately sprinted into a nearby stand of cattails and disappeared, while the other took to the air, careening haphazardly across the pond as if flying were a last resort it had rarely bothered to practice. For its part, the snipe froze in place, and I couldn't help picturing a thought bubble above its head: "Don't see me . . . Don't see me." The fox turned to watch the flying rail, so distracted by that motion that it did indeed miss the snipe completely, trotting past within a few feet and then continuing out of sight into the brush. Soon the snipe was probing in the mud again, but not for long. Its morning meal seemed destined for interruption, as described in this excerpt from my notes:

> **07:28—Snipe** suddenly dashes up shoreline and flight-hops into lawn as raccoon appears, lunging after it. Raccoon chases and snipe takes flight, low over the pond, almost crashing into fir log.[3]
> **07:30—Raccoon** moves on, and snipe returns, landing again on pond's edge. Now chased and harassed by hummingbird. Hummingbird persists, divebombing, and snipe flies off over yard, tiny hummer in pursuit; two fleeting bird shapes.

Over the next twenty-five minutes, I watched with growing amazement as the pond attracted a doe with two fawns, a mother raccoon with two kits, two more lone raccoons (one with a ragged ear, one without), a variety of songbirds, and a swarm of what appeared to be small beetles, twirling in endless circles on a patch of open water beneath the willow tree.

The rails made several more appearances, racing back and forth along the pond edge and calling out to one another with their clicks and snaps and hoarse, descending grunts. Even the snipe came back, making one more attempt at a peaceful breakfast before getting driven off for good by the approach of a curious fawn. Eventually, the animals all dispersed and I went back inside to finish the dishes, but not before jotting a quick conclusion in my notebook: "I have viewed wildlife all over the world, but never experienced a better half-hour than this."

To be clear, our pond is not some pristine wilderness habitat. A previous landowner carved it out with a backhoe, and it still looks pretty rough—shallow and muddy, with the spoils heaped into a haphazard berm on two sides. But after weeks of summer heat with very little rain, it had apparently become the local equivalent of a Serengeti water hole. Was it that busy every morning? Had I been literally "over-looking" a biological hotspot right outside my kitchen window? To find out, I obviously needed to spend more time watching the pond. But I wanted to see it up close, in a way that wouldn't cause a disturbance. Which meant that, if animal vision worked like human vision, I would have to put myself somewhere that their brains would learn to fill in as part of the background. For example, hidden inside a heap of sticks and branches.

Full disclosure: I can't claim this idea as my own. The inspiration for what quickly became known around our household as Papa's Stick Pile traces directly to a previous generation, and the work of a naturalist who was once a household name. In the mid-twentieth century, Edwin Way Teale belonged to an elite circle of famous North American scientists and nature writers. He maintained a decades-long friendship with legendary bird-watcher and field guide innovator Roger Tory Peterson. He corresponded regularly with Rachel Carson, offering encouragement and research materials for her early drafts of *Silent Spring*. His own books about the nature of spring, summer, and autumn had all been bestsellers, and in the early 1960s he was hard at work on a follow-up about winter that would earn him the Pulitzer Prize. Teale had a knack for finding new perspectives on familiar landscapes. He kept an "insect garden," and developed novel techniques for photographing its residents up close. With his wife and frequent collaborator, Nellie Donovan Teale, he pioneered the idea of the natural history road trip, traveling thousands of miles by car to seek out particular species, habitats, and seasonal events. And at home on their farm in Connecticut, Teale found a memorable way to both work on his books and make observations of backyard nature at the same time.

"A visitor once suggested I might call it my branch office," he once wrote, and described his outdoor study: a simple plank seat, a table, and a framework of two-by-fours, "around which are heaped dead branches gathered from the woods."[4] From this hidden perch, Teale peered out through various cracks and gaps while he worked, keeping tabs on all the wildlife passing unknowingly by. He watched catbirds harvesting strips of bark to line their nest, and spotted a chipmunk chasing after a rolling hickory nut. Woodchucks, squirrels, grosbeaks, and green herons all revealed themselves in unscripted, unconscious moments. Teale's brush pile opened "a hundred windows" onto the constant small dramas playing out when backyard animals don't know we're watching.[5] As he put it, "I see without being seen."[6]

I set about replicating Teale's "branch office" on a day in early April, when winter and spring were still arguing over the weather. Warm sunlight alternated with clouds, and a chilly breeze gusted erratically through the trees, streaking the pond with dark ripples. For branches, I had dragged in a heap of windfalls from the woods, just like Teale, but I cut alder poles to build the frame rather than buying lumber. I also used an old deck chair in place of Teale's bench seat, and decided to forgo the idea of a table altogether. Where Teale described his construction as "a beaver lodge on land,"[7] mine ended up looking more like a beaver lean-to, and it became even shakier a few weeks later, when a falling tree partially stove in the roof. There were also discrepancies in setting. Teale sited his lodge where a rushing brook tumbled over a waterfall into a clear pool surrounded by ferns. Mine sat where the muddy overflow from our pond passed through a ditch and drained into a salmonberry thicket. But as soon as I began spending time there, I realized that the lesson of the stick pile had little to do with its appearance or its location. Instead, my hidden bower taught me something relevant to nature-watching anywhere: the power of sitting still.

As a general rule, close-up interactions with wildlife involve some level of startlement. We are surprised when we come upon an animal, the animal is surprised to see us, and there is a brief moment of recognition before one or both parties turn to flee. Occasionally, we might be lucky enough to spot the creature first, and witness a few candid behaviors before it senses our presence. More often, it is the animal that has the advantage, and we see little more than its rapidly departing backside. Overcoming such limitations typically involves watching wildlife from afar through binoculars, or, in some cases, slowly habituating animals to our presence. Early in my career, I devoted two years to doing exactly that with groups of mountain gorillas in Uganda, following them on a daily basis and painstakingly narrowing the distance at which they were comfortable.[8] A far more straightforward option is to simply stay put, out of sight, and spy on whatever wild things happen to pass by. Nature photographers and filmmakers know this trick, and we've grown accustomed to letting them do it for us. But no image or video—however beautiful or informative—can replace the full sensory experience of an in-person encounter, and the thrill of glimpsing other species behave in ways they never would if they knew we were watching. Sitting in my stick pile reminded me of all of that, starting with something that happened before I'd even finished building it.

Opening up a clear view of the pond involved a lot of branch adjustments, and I was checking the latest arrangement from the chair when I noticed something other than wind rippling the water's surface. I froze and watched the shape coalesce into an upturned nose, two rounded ears, and the long, narrow body of a mink. It swam with ease and purpose, four feet paddling and sleek tail trailing out behind like a rudder. I watched as it skirted a half-submerged log and then turned and came directly toward me, so close that it passed out of sight behind the stick wall at my feet. Hardly breathing, I heard the faint purl of fur parting water as the mink emerged on the shore, inches away, and then I caught another glimpse of it through a gap in the branches. It paused on the muddy bank and shook once, doglike, sending up a spray of tiny droplets, before bounding off down the drainage ditch. Not once did the mink look in my direction.

Wildlife watching from the stick pile wasn't always so dramatic. As the weeks played out I would spend many quiet hours there too, where the highlight might be a striped leafhopper perched on my notebook, or a new appreciation for the architecture of stinging nettles—how their paired, symmetrical leaves alternate and diminish up the stem, like the tiered rooflines of a pagoda. With views of the outside world limited to gaps between branches, sounds took on great significance. I suddenly understood why so many of Teale's observations had started with something he heard—the amplified rattle of rodents rummaging through dry leaves, for example, or the crash of a kingfisher diving headfirst into his pool.[9] More than once I found myself aware of black-tailed deer foraging nearby, invisible, but so close I could hear the tug and tear of leaves ripped from twigs, and the steady, ruminant clacking of their teeth. Unseen songbirds often bathed just out of sight along the shrubby shoreline—soft, erratic splashing sounds followed by a sudden flutter as they waved their wing feathers dry. Smells, too, seemed more potent in the stillness of the stick pile, from the wet earthiness of pond water to the spicy perfume of wild roses warmed by sunlight. Sometimes birds and animals paraded by in plain sight; often they didn't. Either way, sitting in the stick pile held the promise of a rich sensory experience.

In addition to letting me spy on the creatures visiting our pond, sitting in the stick house put me on closer terms with its many residents. Rarely did I see the water free from the skittering presence of striders and skimmers, and the air above it hummed all spring and summer with hatch after hatch of mayflies, caddisflies, and more. Tadpoles crowded the shallows, and through binoculars I occasionally spotted the bulging eyes and smooth heads of their parents—adult tree frogs and bullfrogs, floating, half-submerged in a film of duckweed. It was while scanning the surface for frogs that I witnessed something hard to fathom: a small stream of water, jetting straight up from the surface. It happened so fast I nearly missed it, and I waited in vain to see another one. But there had been no mistaking that tiny geyser—a sudden burst, as if some underwater creature had pointed a child's squirt gun up at the sky and pulled the trigger. As it turns out, that's not far from the truth.

I had been thinking about this very phenomenon (and hoping to see it) for several years, ever since a friend spotted something similar in another pond nearby and asked me to help solve the mystery. At the time, my mind immediately turned to archerfish, famous for their ability to spit water up to 10 feet (3 meters) through the air with remarkable accuracy. They use this deadly aim to knock unsuspecting insects from low-hanging branches and leaves, and then scoop up their victims in great predatory gulps as soon as they hit the water. Unfortunately, none of the world's ten archerfish species are known to stray beyond the ponds, streams, and estuaries of Southeast Asia and Northern Australia. To find a spitting pond dweller in North America, I had to investigate a very different group of animals, and seek out an expert on their strange breathing habits.

"A final instar darner would be capable of ejecting that much water!" Phil Matthews enthused in an email. He was referring to the last and largest larval stage of some of the biggest dragonflies on the continent. As to why a young dragonfly would be squirting water in the first place, few people were in a better position to know. For over ten years, Matthews and his students at the University of British Columbia have been studying the many ways that aquatic insects extract life-sustaining oxygen—from the water, from the surface, or sometimes from bubbles of air carried below like tiny scuba tanks. Juvenile dragonflies use a combination of techniques, with a special twist.

Inhabiting ponds and streams for months or even years before emerging as flying adults, young dragonflies are active predators, with leggy, wingless bodies very unlike the wriggling, grublike larvae of many other insects. Entomologists refer to them as "nymphs" or "naiads," though their spiky, segmented abdomens and jutting mouthparts seemed to stretch the comparison to fairy maidens.[10] Dragonfly nymphs breathe by drawing water into their abdomens past a set of highly efficient (if unfortunately named) rectal gills. To put it bluntly, they breathe through their anus. While that may sound awkward, it gives the nymphs an added and unusual ability: movement via jet propulsion. By expelling their breathing water back out again with sudden force, nymphs can lunge toward potential prey with surprising speed and agility. In the Lilliputian world of a pond water food web, dragonfly nymphs rank as apex hunters. But they can also use their jets to dart away from danger,

FIGURE 7.2. When startled at the surface, a dragonfly nymph will sometimes emit a jet of water into the air as it darts below. Illustration © Chris Shields.

and that brings us back to the mysterious squirt I observed from the stick house.

Most of the time, dragonfly nymphs do their breathing and jetting around out of sight underwater, often near the murky bottom. But if dissolved oxygen becomes scarce, as it often does when small ponds heat up in summertime, some species will float inverted near the surface, extending their rear ends toward the thin, oxygen-rich layer where water transitions to air. "If you startle them when they are air-breathing in this way," Matthews explained to me, "with the tip of their abdomen just below the surface of the water, they'll forcefully jet to quickly dive deeper." This almost certainly explained the squirt that I'd seen. According to Matthews, any threat might trigger such a response—a sudden movement, a noise, or a looming shadow passing overhead. His opinion carried weight, because unlike most of us, Matthews understood this behavior from direct personal experience. "I've been shot in the face before," he recalled, "by nymphs doing just this in my experiments."

Metaphorically, it's fitting that something as visible as a simple squirt above the waterline has such an arcane and fascinating explanation below. Time and again, those of us living and breathing in the open air underestimate the complexity of lives aquatic. When an unfamiliar plant appeared on the pond, I examined its tiny, heart-shaped leaves under a microscope, identified it as a floating liverwort (*Ricciocarpos natans*), and assumed the story would end there. Then I spent a rapt hour peering at the incredible diversity of creatures I'd scooped up along with it. There was a tiny, translucent snail and scores of split-tailed larvae feeding on the liverwort's rootlike fringe. A cluster of minute eggs clung to the underside of a leaf, and little transparent rods and zeros filled the water, spinning and turning and pushing themselves forward with thin, flailing tails. Backlit from below, something long and skinny wriggled inside a glassine tube attached to a rootlet. Occasionally, it would lunge out at a passing swimmer, shaking its head and snapping its fierce mouth like a Chinese dragon. I recognized almost nothing, a thrilling sort of ignorance that must have echoed Dutch scientist Antonie van Leeuwenhoek's feelings in 1702, when he first gazed at the magnified roots of some "Green Stuff" gathered from a canal behind his house.[11]

"I was surpris'd with the sight of a great many and different kinds of *Animalcula*," he later wrote, using his own catchall term for microscopic organisms.[12] "Two Sorts had long Tayls," he added, and described how their bell-shaped bodies kept the water

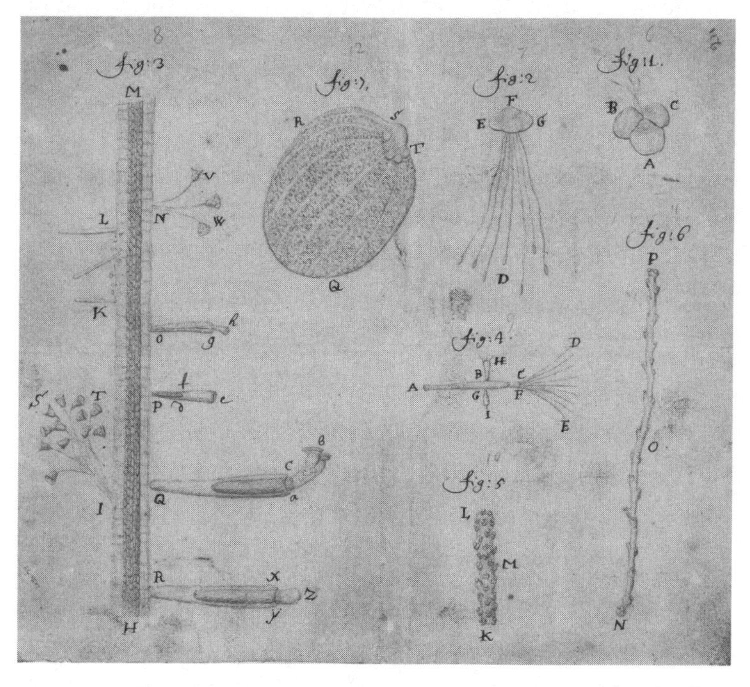

FIGURE 7.3. Dutch microscopist Antonie van Leeuwenhoek commissioned this illustration of duckweed he gathered from an urban canal near his home in 1702. It shows the leaves and roots at various scales, as well as associated microorganisms including ciliates, vorticellids, rotifers, and a hydra. Artist unknown (1702). Image © The Royal Society.

in constant motion.[13] He also spotted "roundish"[14] creatures living in sheaths of various sizes, some with hornlike protuberances and others with tiny retractable wheels, toothed like cogs in a watch, revolving steadily with a wondrous, "swift Gyration."[15] With these careful observations, van Leeuwenhoek added rotifers, hydra, and vorticellids to a list of discoveries that already included such basic forms of life as bacteria and protists. At the time, his groundbreaking microscope techniques (and his refusal to share them) made van Leeuwenhoek perhaps the only person on the planet capable of readily seeing such tiny things, and he is widely celebrated for more or less inventing the field of microbiology. But he should also be an inspiration to anyone curious about the nature in their backyards.

A reluctant traveler, Antonie van Leeuwenhoek rarely strayed far from his birthplace in the small Dutch city of Delft. Nearly all of his major discoveries took place there, either at home or in the meadows, waterways, and woodlands on the edge of town. In addition to the neighborhood canal, van Leeuwenhoek sampled from such exotic locations as the well in his courtyard, and the roof of his house, where he searched for signs of microscopic life in rainwater sluicing through the gutters. When his curiosity turned toward larger creatures, he also focused on species and habitats close at hand, including "some Dragon-flies flying in my Garden."[16] Through his microscope, van Leeuwenhoek counted the hexagonal facets of each compound eye ("more than 8,000"),[17] and then peered through those multiple lenses as a dragonfly would, marveling at how one candle flame was transformed into hundreds, and how a nearby church suddenly boasted a "great many small Towers."[18]

His enthusiasm also led to revelations about dragonfly reproduction. At a time when many people believed that insects generated spontaneously from rotting flesh, van Leeuwenhoek made detailed descriptions of dragonfly sperm, and noted the position of the male and female during copulation. He also produced what must be the first illustration of a young dragonfly, hardly more than a dark ribbon, still curled within the translucent walls of a fertilized egg.

Antonie van Leeuwenhoek made scores of discoveries in places his contemporaries overlooked, and not just because he built better microscopes. Even inferior lenses could find much to comment upon in common places, something English physicist Robert Hooke soon proved by confirming many of van Leeuwenhoek's observations (and adding new ones all his own) using a scope with only a fraction of the magnifying power. In science, sometimes the quality of the tools is less important than what you think to do with them. Few modern examples illustrate that principle better than the advent of environmental DNA.

Biologists have been identifying species by patterns in their genetic code for decades, extracting DNA from hairs or blood samples like forensic experts at a crime scene. Only recently have those techniques been adapted to the masses of DNA found in what amounts to organic detritus—the scraps of bodily tissues and fluids that organisms sluff off as they go about their daily lives.[19] This eDNA accumulates and circulates particularly well in water, which means that a small sample from a pond or river reflects biological activity over a much larger area. With such a tool at his disposal, van Leeuwenhoek could have put a name to all the tiny creatures in his sample of canal water, as well as any minnows swimming around upstream, or a duck that happened to paddle by the day before. Analysis requires a laboratory, but collecting the water is relatively simple. High school students in Nanjing, China, for example, recently collected samples from the city's wetlands that yielded evidence of over nine hundred species, a surprising degree of urban biodiversity that included everything from fungi and bacteria to mollusks, fish, amphibians, reptiles, and birds. Volunteers armed with eDNA collection kits are now a regular part of backyard surveys for great crested newts in England and Wales, and also helped conduct the first national-scale inventory of coastal fishes in Denmark. The tool is also transforming how biologists monitor populations of rare and obscure species that might otherwise require months of fieldwork to locate and identify. Among many examples is the Hine's emerald, an endangered dragonfly in the American Midwest with a life history seemingly designed to evade detection. Rarely spotted during its brief adult flight period, it spends the vast majority of its life as a nymph, eking out a living in calcareous fens, deep in the muddy burrows of a certain species of crayfish (also hard to find). Now, instead of searching the skies above wetlands for a few weeks in summer, or pumping out crayfish burrows in winter, surveyors can look for the Hine's emerald whenever they have a moment to stop by a likely site and scoop up a vial of water. This convenience, incidentally, helps them address another maxim in science: it's not just the tools you use and how you use them that matter, it's when you choose to do it.

Timing makes a big difference in nature, a theme on full display at my stick house on a day in midwinter. Leafless willows and alders had opened up new views in all directions, but, from a biological perspective, there wasn't much to see. At least not compared to summer, when the promise of lush plants, tasty insects, and plentiful water seemed to draw every animal in the neighborhood. But as I settled in to watch—with a lap rug and a hot water bottle to help stay warm—subtle, cold-season details began to emerge. I noticed a few hardy water beetles boating along, adding their small wakes to the wind lines from a southerly breeze. A precocious woodpecker drummed somewhere nearby, as if by calling for a mate he could will an early start to spring. And buzzing wings overhead alerted me to an Anna's hummingbird darting around in the bare alders—what could it possibly be finding to eat up there? Every season held its own surprises and puzzles, at the stick house or anywhere else in our yard. But it suddenly struck me that something was missing from all of my explorations, regardless of where or how I'd been looking. I had a timing problem. Considering *when* I'd been making my observations, it's no exaggeration to say that I was missing half the story.

CHAPTER EIGHT

After Hours

Whoever thinks of going to bed before twelve o'clock is a scoundrel.

—Samuel Johnson
Apophthegms, Sentiments, Opinions and Occasional Reflections (1787)

I felt the wind before the talons. It shifted the hair on the back of my head, an instant of silent downdraft, like the warning of an air raid received too late to mount a defense. Fortunately, the bird veered off at the last second and its claws only grazed me, as if making a small adjustment to my morning hairdo. I flinched and threw my hands up, after the fact, feeling my scalp for scrapes and thinking, *Not again*. Then, peering into the predawn gloom, I spotted the all-too-familiar shape of a young barred owl, perched on a nearby fencepost.

"Goddamn it, Owl," I said under my breath, whispering to avoid waking Eliza and Noah in the tent behind me. The bird blinked impassively, its eyes perfectly round and intensely black in the pale disk of its face. I waved my arms and approached until it took flight, flapping without sound over my head to land on a fir branch directly above the trail to the tent. That would never do. I couldn't just leave it there, ready to pounce on the next sleeper to arise. A well-tossed stick or fir cone might scare the bird off, but there didn't seem to be anything suitable near my feet. When I looked up again the owl was nearly on top of me, a great swooping crescent of outstretched wings, hooked bill, and bad intent, three feet from my face. "Hyaar!" I hissed, and raised my arm, and the bird hovered for a moment directly before me, like a floating partner in some kind of ghostly, formal dance. Twice more we repeated our feathery tango, the owl diving and then retreating to various branches until it finally flew off, out of the trees and across a field beginning to brighten with daylight.

Attacks like this had become commonplace in our yard, particularly near the house and along the path to the tent where we liked to sleep during the summer months. We sometimes wore bicycle helmets or carried umbrellas to protect our heads, and learned to always scan the owl's favorite perches whenever we went outside. It wasn't an idle worry. Barred owls sport wickedly sharp, inch-long (2.5-centimeter) talons and can strike with a force more than fifteen times their bodyweight.[1] They've been known to lacerate unlucky joggers passing near nest sites, and death-by-owl remains a credible theory in one of North Carolina's most famous unsolved murders.[2] (Forensic experts found talon-like wounds on the victim's head, and three tiny feathers entangled in her hair.) The sheer power of their strike helps barred owls bring down a huge variety of prey, including squirrels, rabbits, opossums, bats, woodpeckers, crows, and even other varieties of owl. In our case, however, the owl was not attacking because it was hungry; it did not want to eat us. It wanted us to go away.

FIGURE 8.1. The young barred owl that chose our yard as its territory, and set about trying to drive us away. Image © Thor Hanson.

Territorial behavior is common in a wide range of species, including our own. We feel instinctively connected to, and possessive of, the places we call home. So it can come as a shock to learn that regardless of how long we've been paying the rent or the mortgage, other creatures believe that our property belongs to them. Barred owls are a good example. Like us, they lay claim to distinct territories, displaying what ornithologist Arthur Cleveland Bent called a "remarkable attachment . . . for a favorite locality."[3] Beginning in the 1890s, Bent and several colleagues documented pairs of owls occupying well-defined patches of pine forest in New England for as long as thirty-four years—defending borders, chasing off intruders, and raising up brood after brood of chicks.[4] With adult owls so determined to stay put, it tends to be the juveniles that move around, fanning out across the landscape in search of empty places to call their own. Our owl could have been just such a drifter, or it might have been the same youngster we'd seen hanging around the yard in late spring, newly fledged and still begging food from its parents. Either way, its aggression probably boiled down to inexperience, a lack of clarity about just who or what counted as a territorial rival. The young owl had yet to learn a lesson that its parents, and probably every other nocturnal species on the property, had long since taken for granted: don't worry too much about the people; they disappear after dark.

It's hard to argue the point. During the day my family spends a good deal of time outdoors—gardening, doing chores, feeding the chickens, splitting firewood, or generally just poking around. But from sunset to sunrise, it's safe to say that the level of human activity in our yard—and for that matter the whole neighborhood—pretty much drops to zero. Biologists describe our species as "diurnal." The word means active during the daytime, a Latin antonym of the more familiar term nocturnal. (Creatures that specialize on the half-light of dusk and dawn get the best word of all, "crepuscular.") Yes, we can extend our waking hours through the use of artificial light, but that behavior is generally an indoor phenomenon. Even in urban centers famous for their nightlife, such as "the city that never sleeps" (New York) or "the city that never stops" (Tel Aviv), the time people actually spend *outside* after dark mostly involves scurrying from one well-lit building to another. These habits make it possible for owls and other true night dwellers to live alongside us virtually undetected, starting and ending their shifts during the long dark hours when we are off the clock. While I can't say that I relished being dive-bombed by an angry bird (particularly if it might decide to take up residence for thirty years), every interaction with our barred owl reminded me that no matter how much time I spent investigating backyard biology during the day, I was still oblivious to everything happening after hours. And if the information archived in libraries, academic databases, and the World Wide Web is any indication, I was far from alone.

At the time of writing this paragraph, a simple Google search for the phrase "diurnal biology" turned up 235 million results, nearly twenty times the paltry 12 million hits for "nocturnal biology."[5] Books and peer-reviewed research articles are similarly skewed, a ratio that leaves one implication beyond doubt: biology needs more insomniacs. But even among those of us prone to wakefulness (myself very much included), serious behavioral roadblocks stand in the way of spending time outdoors at night.

Simply put, to one degree or another, everyone is afraid of the dark. It's not a matter of watching too many scary movies, or reading stories about things that go bump in the night. From an evolutionary perspective, fear of darkness has always been a clever adaptation for human survival.

The community of large carnivores in Africa during the Pliocene and Pleistocene spanned at least fifteen genera and scores of species, from familiar felids like leopards and lions to a range of long-gone beasts, including short-faced bears, giant hyenas, and scimitar-toothed cats.[6] Some lived alone, others traveled in packs, but aside from the cheetah, which was (and remains) largely diurnal, all of the prehistoric predators capable of consuming our ancestors did most of their hunting after sundown. In that context, fear of darkness was an advantageous response for early hominins. Those individuals who lacked it and strayed incautiously away from the cave or campfire at night led shorter, riskier lives, with less chance of passing along their reckless genes.[7] Over time, nocturnal cautiousness prevailed, and we all bear the mark of it.

Psychologists know our primal worries well. In study after study, darkness and other atavistic threats (snakes, spiders) trigger stronger and more sustained fear responses in the human body than modern hazards like handguns or busy traffic.[8] The newer threats are arguably more dangerous, but we've known them for only a few generations, not nearly long enough to evolve an innate response. To be sure, our hesitancy to explore outside after dark can be overcome—think of Darwin making midnight worm forays at Down House, or Doug Tallamy patiently counting backyard moths with his light and bedsheet.

But for most of us, it takes a conscious effort, or some kind of external stimulus, to leave the habitual comfort of a well-lighted place. My own nocturnal observations continued to lag long after I had anything to fear from our dive-bombing owl. It was autumn, the owl had moved on, and we had moved our sleeping quarters back inside, except for Noah, who had shifted to a plywood shack on the lawn to extend the outdoor season, and enjoy some teenage independence. That's where he had headed on a chilly evening in October when he suddenly appeared again at the kitchen door.

"Papa, are these your migrating geese?" Noah asked, and I rushed outside to join him in the darkness. It was high season for birds in passage, and we had just been discussing what to listen for. I hadn't expected the payoff to come so soon, but there was no mistaking what we heard from the porch: faint honks and calls descending through the fog like grace notes from a distant choir.

"Snow geese," I confirmed in a whisper, and we grinned at one another in the dimness. "Aren't they great?"

I told him the birds were pure white, with black wingtips and bright orange bills. They passed high above our island every year in great undulating flocks, traveling between their arctic breeding grounds and the farm fields, wetlands, and river deltas where they wintered. We could see nothing through the mist, but it must have been clear a few hundred feet up. The birds need sightlines to the constellations and landmarks that help them navigate, and I could picture them up there, flapping hard in long, pale *V*'s over a pillowy sea of fog and mountaintops, everything bathed in starlight. At night, the thrill of biology sometimes lies less in what you see than in what the darkness forces you to imagine.

Later, when everyone was sleeping, I grabbed a field notebook and a warm jacket, and headed outside to take stock. Droplets of mist glowed in the beam from my headlamp as I made my way slowly up the driveway, and I could still hear sporadic goose calls drifting down from on high. At ground level, life seemed more subdued, hushed by fog and darkness, with the notable exception of tree frogs. They croaked loudly from the vegetation all around me—long, creaking noises that sounded like a slow-motion drawl compared to their fevered chorus in springtime. Male tree frogs in autumn don't sing to perform so much as to flex, giving off throaty growls that some experts interpret as practice, some associate with the frog equivalent of puberty, and others believe to be an innate response to any springlike combination of temperature, moisture, and day length. Regardless of why they do it, how they do it remains remarkable, a combination of vocal vibrations and bodily resonance known to exceed eighty-five decibels at close range. That's roughly the same volume as a lawnmower from a creature the size and weight of a matchbox. So *hearing* frogs call is a simple matter. *Seeing* them call, however, is altogether trickier.

I turned toward the closest-sounding frog and sidled up to the shrubs, switching the light on my headlamp to red. I had learned this trick after many fruitless frog searches around our pond during the breeding season, when, no matter how stealthily I approached, the entire chorus would go silent at the first hint of a flashlight beam. Frogs have exceptional night vision for the colors green and blue, and even into the ultraviolet range, but it turns out that most species can't see red at all. To them, a red light is just another shade of darkness, making it possible to sneak up close and watch their behavior unobserved. Our own eyes work just fine in a red glow, adjusting as they would to any dimness, with widened pupils primed to pick out every possible detail. But it's one thing to find a frog in a spring pond teeming with them, and quite another to spot one in autumn when they are all spread out on land. (Although we associate frogs with water, Pacific tree frogs and many other species only visit ponds during the breeding season.) I began looking carefully through the foliage, examining twigs and turning over leaves in the dim red light. It was painstaking work, but not idle. If I could spot a frog making one of those long, grating croaks, I might be able to help resolve, at least in part, a long-standing mystery in the world of acoustic biology.

"The vocal sac is a surprisingly complex structure," Jeffrey Ethier told me in an email. A doctoral candidate in herpetology at the University of Ottawa, Ethier had recently collaborated on a thorough review of the genus *Pseudacris*, a group that included our tree frogs alongside a range of related species known variously as chorus frogs or peepers. All of them are noisy, and all of them inflate huge air sacs below their throats when they sing. Acoustically, scientists long believed that these sacs acted as resonators, increasing the volume of vocal notes in the same way that a hollow guitar body amplifies the sound of a plucked string. "Undoubtedly, there is some role of resonance," Ethier confirmed, but tests of the theory had proved inconsistent. Experiments with museum specimens found that many sac tissues failed to resonate at the same pitch as their frog's vocal cords, and even those that did produced only modest gains in volume. Alternatively, the sac might serve as what Ethier called a secondary sexual signal, another way for males to convey their health and fitness to potential partners. In some species, the sac is adorned with a bright, contrasting color to accentuate this possible effect. The most promising new theory, however, follows another musical analogy, equating the vocal sacs of frogs to the tautly inflated bladders found on bagpipes.

"The current research suggests that it helps recycle air back into the lungs without having to actively inhale," Ethier explained. This may sound like a minor benefit, but in the context of competing vocally for a mate, it means a lot more noise for a lot less effort. By filling their vocal sacs and calling with their mouths tightly closed, frogs can rapidly pump the same big breath of air back and forth across their vocal cords again and again. Find a frog chorusing in springtime and you can see this process in action, with the tight skin of the sac contracting visibly between ribbits as it forces air back into the lungs. For bagpipers, inflating and pressurizing a bladder makes airflow to the reeds continuous, allowing the production of a steady drone and all the accompanying trills and melodies. Frogs get a similar benefit, using that steady air pressure to sing more rapidly and efficiently than they could if they had to inhale after every croak. (Veterans of high school biology dissections may recall that frogs lack a diaphragm, an evolutionary omission that turns breathing into a multistep process of sucking in and swallowing individual gulps of air.) Biologically, all of this makes

FIGURE 8.2. The inflated vocal sac of a male Pacific tree frog (*Pseudacris regilla*) is thought to help amplify and redirect sound, and act as an air bladder to increase the speed and efficiency of calling. It's also an impressive sight, possibly acting as a visual display to attract females. Image © Mike Benard.

perfect sense for vocalizations during the breeding season, but what about the slower, more sporadic calls of autumn? Do frogs bother to fill their vocal sacs when speed doesn't matter? The answer to that question would either strengthen or undermine the bagpipe theory. Finding that answer, however, required finding a frog. And that was proving difficult.

Minutes passed as I rummaged fruitlessly through the vegetation. Whenever the frog croaked, it sounded close enough to reach out and touch. But it was impossible to pinpoint, reminding me that vocal sacs had yet another proposed function. According to some experts, their round shape acts like a high-end, 360-degree Bluetooth speaker, broadcasting on all sides and making frog calls seem omnidirectional. That, at least, was a theory I could substantiate by failing to locate the frog. In my defense, it's possible that all the frogs I was hearing were hidden inside their hibernacula, shallow depressions under leaf litter or loose soil where—considering the chilly weather—they might already have hunkered down to wait for spring.

Before moving on, I looked up to see if perhaps my ventriloqual quarry was overhead, singing in plain view from a crab apple branch. No such luck, but the glare from my headlamp did trigger a response. Unlike frogs, birds can see the color red perfectly well, and there was a sudden eruption of flapping as my light startled one awake. I heard it lurch away, raking the twigs with agitated wingbeats as it rearranged itself in the canopy of a nearby ocean spray. A sparrow? A thrush? I never did see the bird clearly. Biology by flashlight is an inherently limited enterprise. But it can also be revealing. Narrowing the world to a single beam helps eliminate distractions, allowing a focus that can be hard to achieve in the visual abundance of daylight. You don't see as much, but what you see gets noticed. Once, that principle helped me find the nest of a tropical wren so cryptic its breeding behavior had never been fully described. Spotlighted by chance at night, the nest was obvious: a perfect sphere of rootlets and palm fibers hanging like a bauble from the stem of an epiphyte. But in full daylight, surrounded by the overwhelming greenery of the jungle, that tiny woven ball became virtually invisible. I certainly hadn't seen it, nor, apparently, had the dozens of other biologists constantly passing by below. It took a flashlight beam to reveal something unknown to science hanging above a place very well-known to scientists: the path to the dormitory at La Selva Biological Station in Costa Rica, one of the busiest rainforest research centers in the world.[9]

I kept the wren story top of mind as I continued slowly up the driveway and along our forest trail, familiar ground made new by narrowed glimpses. No cryptic nests appeared, but the flashlight revealed plenty of things that I had been stepping (or occasionally crawling) over in my daily commute. A coral fungus, antlered and pale, plunged upward from the trampled path like an inverted root, and a flush of grassy seedlings ringed the ridged edge of a dry puddle. I noticed how a small fir stump had completely healed over with bark, a sure sign it was still living, grafted by the roots to a neighboring tree. And then I rediscovered a hollow in a rotten log where I had once hidden an Easter egg, its dark entrance now webbed across and guarded by a spider with shining eyes. Most arachnids don't see red light either, and neither do ground beetles, which allowed me to lie down right beside a narrow-collared snail eater and watch it hunt. It lifted its long legs gracefully but with surprising speed as it zigzagged among the fallen leaves and detritus, its haphazard route a classic example of what animal behavioralists delightfully call "path tortuosity."

My own path curved and wandered through the woods until I felt a chill set in and decided to steer a straight course for home. Emerging from the trees, I stood momentarily blinded by the brilliant white light streaming from the house. After an hour or more in darkness, the glare felt almost painful as my eyes struggled to adapt. Squinting, I could see that there was only one light switched on, the LED equivalent of a single 75-watt bulb, hanging in a shaded fixture over the kitchen table. But that was enough to make the house too bright to look at directly, while at the same time wrecking my night vision and turning everything else to black. In that instant of disorientation, I felt a sudden sympathy for the literal deer in the headlights, and all the other creatures bewildered by human illumination. Academically, I had long known that light pollution posed a challenge for a wide variety of species, but I had never before internalized the biology behind their dazzlement.

From an evolutionary perspective, artificial light is just that—artificial. There is nothing remotely like it in nature. The first commercially successful electric bulbs sputtered to life in the laboratories of Thomas Edison and Joseph Swan in 1880.[10] Arc lamps and gas-based systems date to a few decades earlier, and if you throw in the more modest glow of candles, torches, and wicks dipped in animal fat, you can extend the era of artificial lighting back for several thousand years. Organisms sensitive to artificial light, however, originated on an entirely different time frame. Birds trace their beginnings to an offshoot of theropod dinosaurs that thrived roughly 160 million years ago, and mammals go back another 50 million years beyond that. Reptiles and amphibians are even older, dating back more than 300 million years, and insects got their start over 400 million years ago, not long after the arrival of jawless fish. In all that time, with the exception of occasional wildfires, meteor strikes, and volcanic eruptions, the ancestors of living species experienced nothing after sunset to compare with the brightness of modern lighting. Natural selection and other evolutionary forces fine-tuned their habits for a fuller darkness, a nocturnal environment broken only by starlight, and the cyclical phases of the moon, which, even at peak fullness, is orders of magnitude dimmer than the average lightbulb. (To experience the difference, take this book outside and try reading it by moonlight!) Now we have flooded the nighttime world with novel sources of radiance, from skyscrapers to streetlights, stadiums, and suburban strip malls. Even a single bulb burning in a small house on a rural lot is bafflingly intense to most night dwellers. Nothing in their long histories has prepared them for this great global brightening, most of which has occurred in less than one hundred years, an evolutionary eyeblink.[11] Small wonder that moths flail at porch lights, sea turtles flounder at beachfront boardwalks, and migrating birds bash by the thousands into brightly lit buildings. And as if those navigational hazards weren't enough to contend with, research shows that artificial light creates a host of more subtle, but no less important, internal challenges.

"Illumination decreases the ability to produce melatonin, the hormone of the darkness," Sibylle Schroer explained to me in a lengthy email. While that might sound familiar to anyone who takes a melatonin supplement to fall asleep at night, it has much broader implications in nature. Melatonin deficiencies affect other organisms too, upsetting rest and activity patterns, increasing stress levels, suppressing immune responses, and leading to a higher risk for certain cancers. But the impacts of illumination don't stop there. As a light-pollution specialist at the Leibniz Institute of Freshwater Ecology and Inland Fisheries in Berlin, Schroer has studied everything from diatoms to caterpillars. She can quickly rattle off a surprising breadth and range of biological impacts, from lower growth rates in fish to reduced parental care in songbirds to the presence of damaging free radicals in the leaves of trees. With human-driven illumination expected to double in extent and intensity in little more than thirty years,[12] and researchers documenting negative effects on seemingly every organism anyone has thought

FIGURE 8.3. Turning off unnecessary lights can save countless backyard insects like this one, a small gray moth drawn to the brightness streaming from the author's kitchen window. Image © Thor Hanson.

to study, some biologists have called artificial light an underappreciated threat to biodiversity on par with habitat loss or climate change.[13] But unlike those other hazards, so vast they often feel hard to address personally, artificial lighting has a solution that anyone can partake in.

"Light pollution is an environmental pollution which can be stopped completely when turning off the lights," Schroer observed. It's that simple, an opportunity presented to us every time we leave an empty room, or close down our offices, homes, and businesses for the night. Similar logic applies to choosing lamps and fixtures, or deciding how to light up a yard, porch, patio, or driveway. As Schroer put it, all those choices should be governed by one overarching question: Do I need the light? When the answer is yes, she recommends dimmable bulbs and shades that cast the light downward only where it is needed. "The light source should be invisible!" Bulb style matters too—anything in the soft, warm range will do less harm to plants and animals than those with a white or bluish glare.[14] Taking these things into account can transform the backyard nightscape and also, as my own kitchen window would soon demonstrate, provide something vanishingly rare in the world of species conservation: instant gratification.

My eyes had adjusted by the time I reached the house, and I paused to inspect the large bay window where light was pouring out into the yard. Sure enough, I found a tiny gray moth perched on the glass, and also a small sawfly with a reddish body and long antennae in constant motion. The sight of bugs on lit windows or circling porch lights is so familiar we have to remind ourselves that they're doing something they shouldn't be doing.

Here were two insects stunned by the glare, diverted from their normal activities and made suddenly vulnerable to predation and exhaustion. As many as a third of insects attracted to artificial lights are dead by morning, creating what entomologists call the "vacuum cleaner effect," a steady diminishment of local diversity as sensitive species get hoovered out of the surrounding landscape, night after night. Even individuals that manage to fly off don't escape unharmed. They've wasted precious hours from their short adult lives—and every minute counts when you may have only a few days or weeks to disperse and find a mate. Thinking of Sibylle Schroer's advice, I went inside and turned off the kitchen light.[15] Seconds later, back at the window, I saw that both moth and sawfly had fluttered away, back into the yard and a more suitable darkness.

With that modest gesture, my nocturnal expedition was transformed from an exercise in observation into one of action, helping two small inhabitants of our yard survive the night.[16] It was a tangible reminder that backyard biology doesn't stop with seeing and identifying the plants and animals around us. We can also take steps to improve their lives and habitats, and, as the following chapters will demonstrate, turning off a light switch is just the tip of the iceberg.

PART THREE

Restoring

Where there is freedom to experiment there is hope to improve.

—Sir Arthur Thomas Quiller-Couch
On the Art of Reading (1920)

CHAPTER NINE

The Welcome Mat

You may drive out Nature with a pitchfork,
yet she will ever hurry back.

—Horace
Epistles (20 BC)

That should take care of it, I thought, stomping with my heel to pack dirt and gravel around the base of the new signpost. For several years, my efforts to nurture a patch of thimbleberry bushes at the head of our driveway had been repeatedly thwarted. They'd come up on their own, a native species related to the common garden raspberry that would bear similarly delicious fruits if they ever got a chance to mature. But while I'd been careful to trim back the brush and grass around them, their position near the road made them vulnerable to our local county maintenance crew. Twice in a row, just as the canes got tall enough to flower, a passing mower had whacked the whole thicket back to ragged stumps. This season I was determined to prevent that from happening again. With the thimbleberries once more grown robust, leafy, and ready to bloom, I had put up a notice to keep all would-be shrub cutters at bay.

Stepping back to admire my handiwork, I stopped short and found myself doing the sort of cartoonish double take usually reserved for slapstick comedies. Was I losing my mind? The words I'd painted on the sign made no sense—worse than that, they encouraged the very behavior I was trying to stop! It seemed impossible, but there was no denying the instructions now printed beside my precious thimbleberries in large block letters: MOW ON.

If you turn this book upside down and read that phrase again, you will immediately see my mistake.[1] I'm happy to report that once the sign was properly oriented, it served its purpose admirably, and our thimbleberry patch has been thriving ever since. In springtime, we always enjoy its flush of broad green leaves and pale flowers, followed in early summer by a tasty harvest of small crimson fruits. Those rewards were an ample return on investment, but to be honest, I considered them something of a side issue. It wasn't until I spotted a woody burl gnarling one of its canes that my true goal was realized, and I knew that in one fell swoop I had increased the biodiversity of our yard by as many as fourteen species.

In botany, the word gall refers to any lump, swelling, or other abnormal plant growth spurred by the activities of another organism. Usually, that means an insect. *Gall* comes from the Latin term *galla*, or "oak-apple," a direct reference to the large, spherical knobs that often form on the leaves and twigs of oak trees. These and other galls were gathered throughout the ancient world for use in dyeing fabric, making ink, and tanning leather. Many also had medicinal value, mentioned by the likes of Hippocrates and Pliny the Elder as treatments for everything from hangnails to toothaches, not to mention bleeding gums, ear infections, and dysentery. Galls were readily available for purchase in the shops of Herculaneum, just down the road from Pompeii, where a vessel containing over 2,800 oak galls was entombed and preserved by the eruption of Mount Vesuvius. During the Middle Ages, fortune tellers added augury to the checklist of handy uses for galls. Opening one and finding a maggot inside signaled a coming famine, for instance. If the gall contained a spider, pestilence was in the offing. To British author Beatrix Potter, galls from rose briars were known as "robin's pincushions," a popular plaything for the title character of her story *The Tale of Squirrel Nutkin*. All of these examples demonstrate that people had learned to recognize and find uses for plant galls long before they understood how or why they formed. That knowledge came later, much of it from a single person, a scientist who was perhaps better known for his work in a rather different field of study.

To most people, the name Alfred Kinsey brings to mind a pair of groundbreaking and controversial books on human sexuality. With the 1948 publication of *Sexual Behavior in the Human Male*, followed by a companion volume on female behavior five years later, Kinsey firmly established himself as one of the most famous (and infamous) scientists of the twentieth century. To entomologists, on the other hand, the Kinsey Reports were

FIGURE 9.1. These woody burls on the canes of thimbleberry (*Rubus parviflorus*) give home to as many as fourteen species of tiny wasps, from the original gall makers to a surprising range of copycats, freeloaders, and parasites. Image © Thor Hanson.

an unfortunate diversion from an otherwise illustrious career. Before turning to sex research, Kinsey devoted decades of his life to the study of gall wasps, the tiny insects responsible for deforming my thimbleberry canes. It's hard to overstate the scale of his contribution. While the eight thousand interviews that Kinsey personally conducted about sexuality may sound impressive, that's nothing compared to the attention he showered on wasps.[2] From the time he encountered his first gall, on a field trip while studying at Harvard University in 1917, to the end of his last collecting expedition in 1939, Kinsey and his student helpers gathered and processed over 7.5 *million* specimens. Kinsey's wasps now reside at the American Museum of Natural History in New York, where they make up a staggering 40 percent of *the entire insect collection*.[3] The numbers alone are impressive, but Kinsey and his team also studied biology and behavior, rearing hundreds of different species in captivity and opening a window into the fascinating struggle for life taking place inside each and every gall.

Because of Kinsey, and those who have followed in his footsteps, I knew that the woody knuckles on my thimbleberry canes got their start in the springtime, when female gall wasps laid eggs in the growing tissues of the stem.[4] Droplets from their venom glands, injected with the eggs, triggered a complex reaction that experts still don't fully understand. Apparently, gall wasp venom has evolved to mimic plant growth hormones, directing the host—thimbleberry, or otherwise—to begin producing a mass of nutritious pith. The eggs become embedded in that growing lump, so that when they hatch the larvae find themselves sheltered from the elements within what amounts to a tasty ball of food. The gall protects them and feeds them while they mature, pupate, and eventually chew their way out as adults to start the whole cycle anew. If you think this sounds like a pretty good deal for the wasps, natural selection would agree. Thousands of wasp species have mastered the gall-making habit, and it has evolved independently in other insects too, including certain flies and thrips. Up to half of all plants are thought to host galls, not just on their stems but also on leaves, roots, and sometimes flowers. Many are smooth and brown, but they can also be hairy or spiky, and wildly colorful. Some varieties look remarkably similar to caterpillars, and people often mistake the fleshier galls for fruits.[5] But perhaps the best measure of a gall's success lies not in its appearance or in the evolutionary "ingenuity" of the creature that caused it, but in the number of other species that have learned to call it home. A single gall often contains an entire community of insects, and the best way to study them involves one of the most straightforward experiments in biology. It's the same thing that Alfred Kinsey used to do whenever he encountered an unfamiliar specimen: collect it, put it in a jar, and wait.

I followed that protocol precisely on an early spring visit to my thimbleberry patch, snipping off an average-sized gall and taking it back to the Raccoon Shack for further observations. Up close, it didn't look very promising. About the size of my thumb, the gall seemed utterly inert: a weathered lump of grayish cane, speckled with black spots that appeared to be some sort of mildew. But I couldn't find any holes where an insect might have chewed its way out, so presumably whatever wasps had developed inside were still there, waiting for the right moment to emerge. In a jar on the lab bench in my office, that moment came several weeks later. Or I should say that moment began, because what ensued was not a single emergence event but scores of them, producing wave after wave of tiny wasps that kept me checking the contents of my gall jar throughout the spring and well into summer.

The first individuals to make an appearance looked like animate peppercorns, with stout, black bodies and stubby, cellophane wings. Then came a crawling handful of iridescence: three different varieties of metallic, golden-green wasps, some with abdomens slender and blunt-tipped, others concave and sharply pointed. There were wasps with long-stalked waists that made their bodies look like dumbbells, and others with needle-sharp ovipositors angled jauntily upward from their backsides. Even without magnification, the jar held an obvious and surprising diversity of forms, but one thing was conspicuously absent. At the end of the summer, when the gall lay empty and all the specimens had been identified and counted, I realized that they didn't include a single example of the original gall maker. Every wasp in the jar was a parasite.

"The amount of work involved in solving these life histories is rather considerable," Alfred Kinsey wrote in 1920, noting the difficulty of predicting which species might emerge from a given gall, let alone working out what was going on inside.[6] In the case of my thimbleberry specimen, there was no sign of the wasps whose eggs and enzymes had kick-started the whole process the previous spring. All those larvae had been attacked and supplanted by other species, a diverse community of freeloaders who timed their own emergence to target various larval stages of the gall maker. Some devoured the youngest grubs soon after hatching, others attached themselves to older larvae and fed slowly on their fluids, and at least one species used the larvae as a stepping stone—a quick meal to boost early growth before going vegetarian and feeding on the pithy center of the gall itself. There was even a hyperparasite in the mix, a wasp whose evolutionary strategy lay in parasitizing other parasites. To add an additional layer of confusion, many gall wasp larvae grow into adults that don't even resemble their parents. The two body types live in alternate generations—one that reproduces sexually and another that reproduces parthenogenetically, often in a different sort of gall or even on a different host plant. After decades of studying such bizarre reproductive strategies, it's no wonder that Kinsey became a sex researcher. In fact, he may well have found the human mating system rather straightforward by comparison!

In total, more than eighty wasps emerged from my study gall, a rich community of insects that would never have moved into our yard without a thimbleberry thicket to call home. And one patch of berry bushes is just a small beginning. Every property has missing pieces—plants, insects, birds, and animals that once called that landscape home. Some might never return, but a surprising number are lingering just offstage, waiting for us to roll out the welcome mat. After my thimbleberry experiment, I traveled to a dead-end road in western Massachusetts, where one young family has spent more than ten years taking the welcome mat concept to the extreme.

"When we started, this was all grass with one apple tree, one cherry, and five or six tulips," Charley Eiseman told me as we ambled slowly away from the front door.

"There were six tulips," Julia Blyth confirmed, smiling. She walked beside us, bouncing occasionally for the benefit of the baby strapped to her chest. At only two months of age, Eiseman and Blyth's daughter was already a seasoned backyard explorer. Little Ayla rode along peacefully throughout our two-hour tour, bundled inside the hooded pouch of Blyth's heavy winter "mum coat." With scattered snowflakes twirling down around us, I began to envy them both the warm jacket. But if spring still seemed a long way off on that chilly April morning, visiting before peak bloom made one thing perfectly clear: Blyth and Eiseman don't need sunshine and flowers to show off the incredible biodiversity they've nurtured in their yard.

"Our first strategy was to just let it grow and see what comes up," Eiseman explained, recalling the uniform acre (0.4 hectare) of closely mowed lawn that originally surrounded their modest split-level home. Ash saplings, sumac, wild grape, and a range of other species soon obliged, sometimes with such vigor they had to be cut back or pulled out, a process Blyth memorably referred to as "editing." As the lawn's transformation progressed, the couple began targeting patches of bare ground, or anyplace the grass looked thin, tucking in plugs or seeds gleaned from native plant sales. They established a vegetable garden, planted berry bushes, and set out a scattered orchard of fruit trees and nut trees, connecting everything with a maze of lightly mowed paths. The result is a rich and riotous mix that even its owners sometimes have trouble keeping track of. "Did we put these there?" Eiseman asked at one point, peering at a clump of coneflowers, and later, when we came across a lousewort, "Did I plant that?" But for Blyth and Eiseman, backyard habitat restoration isn't so much about growing particular species in particular locations. It's about creating complexity and structure, boosting biodiversity by setting the stage for the drama to begin, and leaving the rest to central casting. Because wherever and whenever a native plant replaces a patch of lawn grass, it attracts its own suite of native wildlife, and if you know what to look for, you can find their stories written on the leaves.

At 2,308 pages, *Leafminers of North America* is by far the longest field guide in my collection. It covers an astonishing array of moths, midges, sawflies, and other insects united by the feeding habits of their offspring. The adults all lay eggs on stems and leaves, where their larvae develop by tunneling through the soft tissues beneath the surface, feeding along the way and leaving behind wild and often diagnostic patterns of lines, loops, and hollowed-out chambers. Look carefully at almost any plant and you should find some, which is probably why *Leafminers* is also the only book I know of that comes with regular updates from its author. "I want the guide to be complete," Charley Eiseman explains on his website, where the book is sold by subscription, arriving in a series of massive digital files. But even after twelve years of work on the project, he's still finding new things to add "on an almost daily basis." Some discoveries come from expeditions and road trips to far-flung locations (thirty-six states and two Canadian provinces to date). Others are sent by correspondents or gleaned from photos uploaded to iNaturalist and a site called BugGuide .net. But no small number of additions to the book have come from Eiseman and Blyth's own property, including one from the very first native wildflower to appear in the lawn after they "liberated" it from mowing.

Daisy fleabane resembles its namesake (the flower, not the flea), a vigorous, narrow-leaved annual with blossoms made up of delicate white rays surrounding bright yellow disks.[7] Plant guides describe it as common in pastures and "low maintenance turf areas," so it's not at all surprising that it appeared in Eiseman and Blyth's newly liberated lawn. What *is* surprising is how that one individual plant, isolated in an acre of grass, immediately attracted three native leaf miners—a fly and two moths.[8] Eiseman recognized one of the moths from its mine, a swollen, hollowed-out area near the leaf tip that he had seen on other species of fleabane. The second moth was harder, but he eventually narrowed it down to a "taxonomically difficult" group of twirler moths. The fly, on the other hand, didn't match up with anything, and a specialist quickly confirmed it as something new. Eiseman tells this story with a sense of awe undiminished by the passage of time (or the thousands of pages he has written about leaf miners). "The very first plant that came up in our yard had a species unknown to science," he said in a recent online lecture, laughing with obvious wonder. "I think that's kind of amazing."

As we slowly worked our way counterclockwise around the house, it seemed that every tree, shrub, and wildflower in Eiseman and Blyth's yard could tell a similar story of rapid colonization. It was as if leaf miners had the place under constant surveillance, just waiting for new plants to appear. In a sense, that's true. Blyth described them as aerial plankton, tiny creatures swept across wide areas on currents of wind, sensing from afar the chemical cues given off by their host plants. The result is a living embodiment of a slightly altered maxim, "If you *plant*

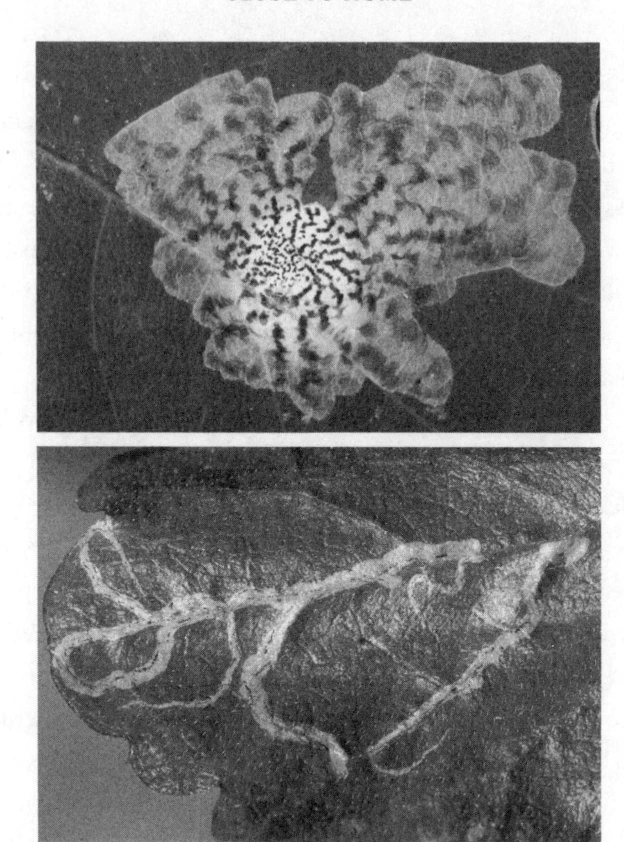

FIGURE 9.2. Leaf miners attracted to native plants can quickly increase the biodiversity of any yard, like these examples from the "liberated" lawn of Charley Eiseman and Julia Blyth. The top photo features a striking pattern left by the larva of a moth (*Tischeria quercitella*) on a red oak leaf, while the bottom photo shows tracks on a daisy fleabane made by what turned out to be a new species of fly (*Phytomyza erigeronis*). Images © Charley Eiseman.

it, they will come." During the pandemic lockdown in 2020, the last time Eiseman had a chance to tally up his and Blyth's finds, the list of leaf miners in the yard had reached 212 species.[9] And that number doesn't begin to account for the wealth of other creatures drawn to this sudden uptick in biodiversity. Native plants, and the insects they attract, anchor the entire backyard food web, providing fodder for everything from spiders and centipedes to warblers and wild turkeys. Eiseman and Blyth have now spotted hundreds of species in their yard that never would have visited a mowed lawn, and even on a chilly morning, signs of that rich diversity were all around us.

"What's that?" I asked as we passed a thicket, pointing to a brownish lump on an old goldenrod cane.

"A goldenrod fly gall!" Eiseman quickly identified, and then he and Blyth searched through other canes nearby until they could show me the slightly different gall shape caused by a goldenrod moth, as well as the neat holes chiseled into the galls by a downy woodpecker eager to feast on the grubs hidden within. The next find was a cane neatly snipped off by the larva of a plume moth. The beveled edge of the cut was distinctive, Eiseman explained, as well as the frass-filled hole where the larva had tunneled downward through the pith of the stem. "They overwinter in the roots," he told me, something he'd figured out by digging up an entire plant, putting it in a plastic bag, and then checking on it daily for a month until the adult moth finally emerged. Nearly forty other species had also come out of the roots, stems, and soil in the bag, including spiders, leafhoppers, springtails, woodlice, and a tiny parasitic wasp with antennae branched like the antlers of a deer. Eiseman had patiently documented and photographed everything, a long-standing habit after spending years rearing thousands of various leaf miners in captivity. Like most insects, the species that burrow into leaves are hard to identify as larvae, so the best way to put a name on them involves the old trick of keeping them in jars until they mature. It's an essential step in the ongoing expansion of Eiseman's field guide, and he always has scores of samples lined up on shelves in his home office. And as if on cue, one of those specimens emerged from its leafy cocoon during my visit.

We found it when I asked to see the place where he had turned so many backyard observations into published scientific outputs. (In addition to the leaf miner project, Eiseman has coauthored a general guide to insect tracks, maintains a popular blog, and churns out a steady flow of peer-reviewed journal articles.) At first, he seemed sheepish about showing me the office, with its leaning piles of books, papers, boxes, nets, cameras, and other signs of a busy biological practice. But his demeanor changed entirely when he spotted something pacing back and forth inside one of his many specimen jars—a shiny, black sawfly with gold-trimmed legs and long, constantly twitching antennae.

Eiseman eagerly snatched up the jar and squinted at a series of letters and numbers penciled on the lid. "I can't read my writing," he said, and then, "Oh! This is from a black cherry leaf. I found it in the yard." He went on to explain how this species rolled leaves into a tube as it fed, rather than mining through them, and that it would find a home in a new book he was working on dedicated specifically to sawflies, a diverse group of plant-feeders related to wasps, bees, and ants.[10] Later that afternoon, the specimen would be processed in what has become a part of the family's regular routine (though he admitted to taking a few days off after Ayla was born). Eiseman would attend to the microscope work and photography, but if the sawfly needed to be spread and pinned for posterity, that task fell to Blyth. She had perfected all manner of preservation techniques as collections manager for the Maria Mitchell Natural Science Museum on Nantucket Island. (The couple met there when Eiseman visited for a conference.) "I don't know how she does it," he marveled, showing me a box of Blyth's perfectly prepared micromoths. Each one was no larger than a fingernail, and all were immaculately arranged with their delicate wings splayed wide, as if frozen in midflight. This was another project in progress—updating the taxonomy of a difficult moth family that hadn't been studied closely since the early 1970s. Eiseman said he was starting to figure them out, but every specimen was a challenge because they were so tiny and looked so strikingly similar. Telling one species from another, he explained, usually required a "full dissection of the genitalia."

Over a lunch of eggs from their own hens, paired with pickled asparagus from the garden, I heard about Blyth's work in permaculture and Eiseman's graduate degree in botany. But it came as a surprise to learn that neither one of them had ever trained as an entomologist. "I took one undergraduate course," Eiseman offered, but the rest—from the leaf miners to the gall flies to the micromoths—was self-taught. And much of that education took place right in their backyard. "When it's in its full glory," Eiseman said at one point, "there's really no reason to go anyplace else."

I had said my goodbyes and was backing out of the driveway when I spotted Eiseman poking around in the yard again, carrying a bulky camera. He was looking for a small gray moth we'd found on our tour, perched low to the ground on a dried stem. It resembled something he knew from a family of leaf rollers, and if he could get a clear picture it might make a good topic for a blog post. "That moth waited for me to come back with my macro lens," he told me later in an email, but on closer inspection it had turned out to be something else entirely. "*Bibarrambla allenella*," he wrote, from a group whose caterpillars use silk to tie leaves into protective, tentlike shelters. "A great genus name," he added, "and not one I've encountered before." Just like that—another new addition for Eiseman and Blyth's ever-expanding "yard list." But not every insect species drawn to recovering lawns requires a macro lens (or dissected genitalia) to be identified. All across eastern North America, people are planting something in their yards to attract—and help save—what may be the most widely recognized leaf-eating creature on the planet.

"Everyone loves monarchs!" Emily Geest declared, and I could see through the video link that her colleagues, Candice Rennels and Emma Webb, were nodding and smiling in agreement. All three of them worked for the Oklahoma City Zoo, a facility situated in the heart of its namesake city and state, smack-dab in the middle of the eastern flyway for monarch butterflies. With their brilliant orange wings trimmed and bedecked in black, monarchs look like animate panes of stained glass, and their biological backstory is just as striking. They mass by the millions in a handful of Mexican pine groves all winter, and then stream northward every springtime in a leapfrogging migration that spans four generations and 2,500 miles (4,000 kilometers).[11] Oklahoma sees them coming and going, hosting the northbound second generation in the spring, and then the southbound fourth and occasional fifth generations in the fall. How the impulse to keep moving passes from adults through eggs and larvae to the next flying cohort remains unclear, but one thing is certain: the whole system depends upon milkweed. Monarch caterpillars will eat nothing else, so completing a successful migration requires a lot of it, in the right locations, all along the butterflies' route. Because of that tight relationship, people in places like Oklahoma are often encouraged to plant milkweeds in their yards, neighborhood parks, and community gardens. Geest, Rennels, and Webb all work on these sorts of monarch conservation efforts for the zoo, including outreach to landowners, collaborations with local schools, and extensive, monarch-friendly landscaping at the zoo itself. But Geest's postdoctoral fellowship goes even further, investigating whether or not such activities actually help monarchs feed, reproduce, and continue on their journey. It's a fitting project for Geest, since it was the sight (and sound) of a monarch caterpillar in an urban setting that got her interested in butterflies in the first place.

"I heard a crunching, chewing sound," Geest said, recalling what would turn out to be a life-changing walk through a park in Omaha, Nebraska, nine years earlier. At the time, she was two weeks into a graduate program in ornithology at a nearby university, focusing on the effects of diseases like avian malaria. But that plan evaporated in an instant when she glanced toward the noise and caught her first-ever glimpse of a monarch caterpillar, wildly striped in black, yellow, and white,

FIGURE 9.3. The sight of this monarch caterpillar feeding on milkweed leaves in a park in Omaha, Nebraska, stopped Emily Geest in her tracks. She snapped this photograph, showed it to her graduate advisors, changed her thesis topic from birds to butterflies, and has been studying monarchs ever since. Image © Emily Geest.

munching away on its favorite plant. "This is it!" she thought to herself, before she even knew what it was. "This is what I want to study!" To their infinite credit, her thesis advisors acquiesced, and a career in butterfly biology was born.

"Monarchs have diseases too," Geest explained, pointing out that the leap from bird malaria to butterflies wasn't as dramatic as it sounded. But the urban setting of her research soon brought up new questions. She found herself interested not only in *how* monarchs got sick, but *where*. Were individuals living in parks and yards as healthy as those in natural areas? Did they survive and reproduce at the same rate? The monarchs migrating through eastern North America had recently been listed as endangered, and anyone interested in helping them viewed such questions as more than academic. Recovery strategies included public campaigns to plant milkweeds throughout the flyway, including in towns and cities. But if those places were dangerous for the butterflies, they risked becoming what ecologists call population sinks, inadvertently reducing monarch numbers by luring healthy individuals to unsuitable sites. To explore that question with hard data, Geest began comparing the fates of monarchs in natural prairies to those she found living in fifteen Monarch Waystations, private and public gardens where citizen scientists had been planting and maintaining milkweeds.

Success in scientific research often depends upon finding differences between the treatments or situations under consideration. For Geest, it was a great relief that everything came out the same. "That's what we were hoping for," she admitted, noting that milkweeds in waystations attracted monarchs at the same rate as those in natural areas, and that rates of egg laying, disease risk, parasitism, and survival were also on par. In other words, the butterflies appeared to be using yards and gardens in much the same way they were using native prairies, an important consideration in a landscape where native prairies had been almost entirely replaced by crops, towns, roads, pastures, and other highly altered environments. Geest was quick to point out that her results shouldn't be extrapolated to all parts of the monarch flyway, and that the quality of butterfly gardens varied widely. (She was currently studying different arrangements of milkweeds and nectar sources, trying to determine the best combinations to plant.) But it was clear that backyard habitats had an important role to play in helping monarch butterflies, and Geest told me she expected that role to grow in the decades ahead.

"Gardens are unique," she said, and began reeling off all the features that set them apart. "They are manicured and cared for. They get watered. They have flowers. They become little islands of recovery, little havens and sanctuaries during droughts." That last comment brought me up short, and I asked her to clarify. Was she saying that monarchs shifted from native prairies to gardens and yards during dry spells? "Yes, during floods too," she answered, and mentioned how recent storms in Oklahoma had overtopped many rivers and creeks, inundating natural areas and driving the butterflies elsewhere. This was an aspect of backyard biology that I had never considered: in an era of climate change, with extreme weather putting huge stress on natural habitats, the patches of ground that we tend around our houses can turn into places of refuge. That idea has significant implications because, as shown by any number of examples in this book (or, for that matter, the briefest of explorations outdoors), our yards and neighborhoods give home to a lot more things than butterflies.[12]

"Monarchs are an umbrella species," said Emma Webb, employing a term that conservation biologists usually reserve for big mammals with large-scale habitat requirements. Saving wolves or bears, for example, involves setting aside wide swaths of land that benefit countless less charismatic species under the same umbrella. In a backyard setting, Webb argued, monarchs achieve something similar. Many other creatures are drawn to the milkweeds and nectar sources planted for monarchs, including other butterflies and dozens of different native bees. There is also a parasitic tachinid fly that arrives specifically to prey upon the monarchs, laying its eggs upon unsuspecting caterpillars so that its larvae can burrow inside and eat them alive.

"The flies are a little harder to get people to rally around," Candice Rennels admitted dryly. She and Webb both work in public relations at the zoo, where they have come to rely on monarchs as vital ambassadors in advancing what Rennels called the organization's two overarching goals: "Connecting people with wildlife and connecting them to wild places." Visit the zoo's website and you will see a beautiful close-up photograph of a monarch on the home page, cycling through in a brief slideshow filled with more typical zoo stalwarts, including a lion, an elephant, and a rhinoceros. The monarch is the only insect featured, and the only species from North America. It is also the only one that zoo visitors can expect to see in their own backyards and neighborhoods, a subtle reminder that you don't have to go on safari to get close to nature. The habitats closest to home are wilder than we think.

Every addition to a backyard species list is deeply satisfying, and when enough people re-wild even small patches of land, the results can be surprising. Across many rural-to-urban gradients, wildlife abundance now peaks in the suburbs, where the sheer variety of habitats is not only richer than that found in the city center, it also exceeds the surrounding, mono-cropped farmlands. And as Emily Geest's research shows, local efforts to boost biodiversity often scale up, helping entire populations of the plants and animals involved. But restoration can do more than attract new species to our yards. It's also possible to take steps that help the plants and animals already living there. Doing so requires a working knowledge of the resources they need to thrive, particularly the things in shortest supply.

CHAPTER TEN

The Limiting Factor

Every limit is a beginning as well as an ending.

—George Eliot
Middlemarch (1871)

In our neighborhood, the arrival of spring can seem like a matter of opinion. Tree frogs feel it soon after the days start to lengthen, sometimes arriving at their breeding ponds when snow still lingers on the ground. The red-flowering currants aren't far behind, leafing out the moment temperatures begin to warm, and then living up to their name with an early flush of bright blossoms. Birds vary in their spring habits, but few are more optimistic than the brown creeper, a tiny woodland dweller for whom any sunny day in February is worthy of full-throated song. Bird books describe the creeper's voice as "thin" and "sibilant," but it might as well be a fanfare of trumpets when the forest is otherwise hushed with winter calm.

This cheerful music greeted me one frosty morning on my walk to work, where the path to my office passes through a dense stand of trees. Stopping to listen, I soon spotted the singer, a lively ball of brindled plumes hopping upward, ever upward, along the trunk of a tall fir. Creepers forage almost exclusively in this manner, ascending the boles of trees and using their curved bills to probe under lichens or flakes of bark, uncovering, in the words of one early observer, "weevils, leaf beetles, flat-bugs, jumping plant lice, leaf hoppers, scale insects, eggs of katydids, ants, other small hymenoptera, sawflies, moths, caterpillars, cocoons of the leaf skeletonizers (*Bucculatrix*), pupae of the codling moth, spiders, and pseudoscorpions."[1] I lost sight of the bird amongst the branches of the canopy, but it soon reappeared, fluttering down to the base of another tree to start the whole process over again. Experience told me to make the most of my creeper viewing opportunities before springtime really got underway. If past was prelude, then this individual and any other creepers hanging around our neighborhood would vanish as soon as it was time to get serious about mating season. Their departure had nothing to do with migration; brown creepers reside in the Pacific Northwest year-round. Nor was it a matter of secrecy; creepers remain bold and easy to observe even when actively nesting. Instead, the birds would leave our woods to overcome a basic scientific principle with an eclectic pedigree. Ecologists and biologists borrowed it from the work of agronomists and soil scientists, but the concept was originally developed by a botanist, before being stolen and popularized by the chemist for whom it was eventually named.

Liebig's law of the minimum states that growth is controlled not by abundance, but by scarcity. In other words, individuals and entire populations are held in check by whatever resource is in the shortest supply, regardless of how plentiful other resources may be. The idea traces back to German botanist Carl Sprengel, who noted in the 1820s that boosting plant growth required identifying and increasing the scarcest mineral nutrient in the soil. Justus von Liebig later passed this theory off as his own, gaining fame for its profound influence on the development of chemical fertilizers.[2] Other branches of science applied "Liebig's" law more broadly, and began identifying any key environmental restraint as a "limiting factor." Predators might be limited by the availability of prey, for example, or herbivores might be limited by suitable vegetation, and the vegetation limited by rainfall. In our woods, brown creepers faced a different kind of limiting factor. They had plenty of tree trunks to forage on, and no shortage of the myriad bugs and spiders they liked to eat. The scarce resource preventing them from breeding and residing year-round was more subtle, but no less consequential: search as they might, they would find no place to build a nest.

In 1879, American ornithologist Thomas Mayo Brewer decided to set the record straight on several points about brown creepers that were "still involved in obscurity."[3] He quickly cleared up questions about the northern extent of their range, and the size of the purplish brown blotches on their eggs. Then he turned his attention to finally settling the mystery of their

FIGURE 10.1. The peculiar nesting habits of the brown creeper (*Certhia americana*) were still unknown to science when John James Audubon composed this painting of the species in the 1830s. Dover Publications.

nests. Although previous authorities, including Brewer himself, had suggested that creepers nested in abandoned woodpecker holes, no reliable account of that practice had ever been published. Brewer called it a legend, and offered instead his observation of a nest tucked obscurely "between the detached bark and the trunk of a large tree."[4] He then cited six other reports of creeper nests "in just such situations and no other."[5] This secretive habit, he concluded, was probably "our Creeper's most usual mode of nesting," and helped explain why their nests had gone undetected and undescribed for so long.[6] Ironically, it also makes them relatively easy to find.

Once you know where to look, spotting a brown creeper nest is simply a matter of locating the right kind of bark: loose, but still attached on top, and cantilevered outward from the trunk to form a narrow gap. Not all such locations will have one, but your odds are a lot better than if you were searching for a typical cup nest, and had to examine every forked branch and twig in the forest. Suitable creeper bark occurs almost exclusively on dead or dying trees, which was precisely what was missing from our yard. Simply put, our forest was too young and vigorous. There was only one sizable dead fir on the whole property, and its bark had yet to loosen. Barring a fire or an outbreak of disease, it would take decades or even centuries to accumulate the mix of healthy trees and snags that make good creeper nesting habitat. If I wanted creepers to stay in the yard year-round, I would have to overcome this obvious limiting factor. It's the sort of challenge that wildlife managers face all the time—adjusting key habitat features the same way a farmer might tweak essential nutrients in the soil. To girdle our beautiful trees and wait for their bark to peel was out of the question, but I had a shortcut in mind, and all that it required was a quick visit to the woodpile.

When you strike a piece of well-cured Douglas fir with a splitting axe, the bark often separates from the wood, much as it does on dead trees in the forest. In either case, the cambium layer has become desiccated and shrunken, creating a natural weak point. For anyone who splits several cords of Douglas fir firewood every year, as I do, the upshot is a constant supply of loose bark in various widths and sizes. Not long after my late-winter creeper sighting, I returned to the same place in our woods and fitted three nearby trees with extra pieces of bark, curved to fit their trunks, and splayed outward at the bottom to create what I hoped were inviting nest sites. It only took a few minutes to set up, and seemed like a worthwhile experiment. Could I entice brown creepers into nesting with such a simple ruse?

To be honest, I didn't have high hopes. Specialized habitat requirements are typically just that—highly specialized, and hard to replicate. Creepers might be cuing in on any number of details unique to naturally loose and peeling bark—color, texture, the size of the cavity, or even something about the overall appearance of a dead tree. My tacked-on slabs of fir bark looked crude, even to my eye, so it came as no surprise to find them unoccupied week after week, as winter passed unambiguously into spring. I had more or less given up on the project when I decided to look again on a warm day in early June. The first little shelter held nothing, as expected, but I was stunned to see a clump of stringy lichen and several small twigs wedged into the second. I tried to temper my excitement—a wren might have placed them, for example, or even a deer mouse. But there was no room for doubt at the third tree, where I found the gap behind my bark stuffed to capacity with an architectural jumble entirely distinctive to brown creepers.[7]

FIGURE 10.2. Overcoming backyard habitat limitations can be surprisingly simple. Here, brown creepers successfully nested behind an extra piece of bark attached to the trunk of a fir tree. Image © Thor Hanson.

"The cavity behind the bark is nearly filled by the nest,—a mass of sticks, bits of bark and dead wood, caterpillar webbing, dry grass, cocoons and the down of cinnamon fern."[8] So wrote ornithologist Winsor Tyler about a creeper nest he found near Lexington, Massachusetts, in 1913. The description holds up well, though in my case the fern down had been replaced by fluff from a willow tree blooming beside our pond. I reached inside the nest and felt the soft, perfectly formed cup where the eggs would sit, and then quickly backed away from the tree to watch from a discreet distance. It didn't take long for the creepers to appear—two subtle flashes of brown, climbing on nearby trunks and fluttering occasionally into the nest and out again, making imperceptible, last-minute adjustments. Both the male and female were attentive parents in the weeks ahead, taking turns incubating the eggs, foraging for the hatchlings, and fastidiously tidying their nest by removing beakful after beakful of droppings—little white gobs of gelatinous chick poop that textbooks primly refer to as fecal sacs. The pace of feeding increased until, on a morning in mid-July, I timed them entering the nest with food every two minutes, bearing insects as large as crane flies and a tiger moth. Choosing a moment when both adults were out hunting, I dared a quick glance inside and saw two perfect chicks, fully feathered, clinging upright to the trunk in classic creeper fashion. They kept absolutely still and silent under my curious gaze, staring back from the shadows with dark, unblinking eyes.

Two days later, the young creepers had fledged and gone, leaving behind an empty nest and a pressing question about our yard: Were there limiting factors for other species that might also be easily remedied? One clear answer took me right back to the issue of dead trees. In addition to brown creepers, at least seventeen other local birds required old snags during the breeding season, from swallows to screech owls, woodpeckers, wood ducks, nuthatches, and more. But where creepers nested on the outside of dead or wounded trunks, all the others nested on the *inside*. Fortunately, mimicking that kind of habitat was even easier than tacking up strips of bark. Instead of visiting the woodpile, I made a trip four miles down the road to a local business that I had been passing by almost daily for years.

Homes for sale. So read a hand-painted sign that stood just outside the main village on our small, rural island. It adorned a roadside stand festooned with the products in question: bird boxes, in all shapes, colors, styles, and sizes. I parked and ambled over to inspect the display, drawn to a peaked-roof variety made from rough-cut boards decorated with horseshoes and curly willow branches. Up close, I noticed how the entry hole on every box was carefully covered with wire mesh, a safeguard to fend off the many prospective tenants I could hear twittering and trilling in the woods all around. (Presumably, it's hard to sell a bird box if it's already filled with birds.)

"You just missed the swallows," said a voice behind me, low and rough, but full of warmth. I turned and found myself face-to-face with the establishment's proprietor, Joe Buckler. A large man with long hair hanging loose, he was leaning out of the open window of his woodshop, gesturing at a scattering of feathers and straw spread across the yard. "They were just here," he went on, and told me how he liked to put out nesting materials and watch the birds swoop down to pluck them from the ground in midflight. He kept several of his nest boxes unobstructed and attached to nearby trees, where various species of swallows and wrens often competed to fill them. Between that ongoing ruckus and the constant stream of sparrows, chickadees, flickers, and other visitors flocking around his feeders, Buckler had plenty of avian activity to keep track of when he wasn't building and selling his boxes.

I introduced myself and we immediately fell into a conversation about the untapped potential of backyard nesting habitat. I told him about the two creeper chicks, and we mused about how many thousands of baby birds his boxes must have ushered into the world over the years—quite a legacy for someone who fell into the business as a last resort. "I broke my back seven years ago," Buckler said, and described how a work injury had cut short his career as a card-carrying union carpenter. Instead of traveling the country from one jobsite to the next, he suddenly ended up stuck at home, unable to do heavy work. Building nest boxes kept his hands busy, and when he put the first few out beside the road, they quickly sold. So he just kept making more. "I wanted to do something positive," Buckler said, and providing habitat for birds felt like a good fit. To cap it off, he crafts each one from waste lumber dropped off by friends still working in the construction industry. "Old fencing, siding from a teardown, siding from a remodel . . ." He nodded at the rows of finished boxes. "All of this would have just gone in a landfill."

From a biological perspective, Joe Buckler is a lot like a woodpecker.[9] He builds what amount to tree cavities, the nesting hollows hammered into deadwood by the sharp bills of nearly every member of the woodpecker family. (Among 239 species, the few exceptions include three that nest in the ground, two that use cacti, one that prefers bamboo, and two that dig into ant nests or termite mounds.) Just like woodpecker holes, Buckler's boxes can be used over and over again by a wide range of

FIGURE 10.3. By his own calculation, retired carpenter Joe Buckler has put over eight hundred bird boxes out into the world, helping cavity nesters from bluebirds to chickadees to owls overcome the scarcity of natural nest sites in most modern landscapes. Image © Thor Hanson.

species. But unlike woodpecker holes, birdhouses are portable, making it possible to add the equivalent of empty tree cavities to any backyard or other landscape that lacks them. Buckler urged me to be experimental. If a box didn't get used, move it, and don't get too hung up on dimensions and appearance. He told me that he tweaks his designs all the time, adapting to the various recycled materials his buddies drop off. "No two boxes are the same," he observed, yet they all find willing occupants. "The birds aren't out there with tape measures!"

Eventually, I settled on four midsized examples from Buckler's inventory, and watched him remove the mesh across their openings before loading them into the car. It was a good start, but I had a hunch it would take a lot more to saturate the nesting potential on our land. "I have one repeat customer who has purchased fourteen boxes," he told me, as if reading my thoughts. "She put one on every post in her garden—no more insect damage!" His comment captured the essence of nesting habitat as a limiting factor: birds would keep filling up additional boxes until some other resource (e.g., tasty garden insects) ran low and became the new constraint on their growth. Considering that our yard contained woods, fields, wetlands, a lawn, *and* a garden, all brimming with bugs and lacking tree cavities, it stood to reason that I might become a good customer too. But I got the feeling that money wasn't the only thing keeping Joe Buckler in the nest box business. (If it were, he never would have given me so many tips on how to build my own boxes.) "Stop by anytime," he told me in parting. "We'll bullshit about birds."

I plan to take Buckler up on that offer, and when I do, we'll have a lot to talk about. Ever since my visit to his roadside stand, I've been adding nesting structures to our yard at a steady clip. The current tally stands at sixteen bark slabs for creepers, twelve wooden boxes sized for everything from nuthatches to screech owls, three hanging songbird baskets, a repurposed mailbox, and one large gourd, hollowed out and fitted with a hole just wide enough for a chickadee. Occupancy rates vary, but few of the nests have gone unnoticed, and certain high-traffic sites attract fierce competition every year. As I write this paragraph, on a mild day in June, our various avian abodes are playing host to the following families: tree swallows (3), violet-green swallows (2), house wrens (2), brown creepers (2), Bewick's wrens (1), and chestnut-backed chickadees (1). For all of these species, the addition of nest sites transformed our yard from fly-through country into good-quality breeding habitat. And birds aren't the only creatures out there searching for some facsimile of a dead tree. Bats also enjoy the shelter provided by peeling bark or hollow trunks, and they wasted little time occupying the artificial roost sites we provided, both intentionally (two bat boxes) and inadvertently (a gap between the rafters above the porch; under the loose siding on the Raccoon Shack). Various solitary bees and wasps also target deadwood, seeking out the tunnels left behind by burrowing beetles. Bamboo nesting tubes and holes drilled into scrap lumber have replicated that niche nicely, boosting our populations of mason bees, mason wasps, leafcutter bees, and more. The same basic concept applies to any limiting factor that can be identified or surmised. Wetting a patch of soil, for example, provides mud for the nests of robins, barn swallows, and a wide range of insects. Adding wildflowers boosts the food supply for pollinators, and a simple sheet of plywood, left flush on the ground, gives vital cover for everything from beetles to snakes to voles. Opportunities for improving and expanding backyard habitats abound—the biggest limiting factor is often our own time and motivation. But if the results streaming in from recent studies are any indication, it's well worth the effort.

Data from a single yard can only tell us so much. It's impossible to know, for example, whether my bark experiment increased the world's brown creeper population by two, or whether those chicks would have simply been raised by their parents in some other patch of forest with a better supply of dead trees. The real power and promise of backyard biology only come into focus at a wider scale, when scientists analyze impacts at the level of neighborhoods, landscapes, or even entire regions. Few examples offer a more dramatic and hopeful demonstration of that principle than a decades-long effort in Switzerland, where landowners and local authorities have cooperated to reverse population declines in one of the world's most threatened groups of organisms.

"In restoration, lots of people do something, but few people check to see if it works." So began my conversation with Swiss herpetologist Benedikt Schmidt, who leads a research group focused on the conservation biology of amphibians at the University of Zurich. I had called to ask him about the origins of a recently published study that boasted some pretty eye-popping numbers. Most field research involves two or three seasons of data collection on a single species at a handful of locations. In contrast, Schmidt and his colleagues had analyzed twenty years of population counts for twelve different species of frogs, toads, and newts in 856 ponds, scattered across the entire 542 square miles (1,404 square kilometers) of the Swiss canton of Aargau. That location made all the difference, Schmidt explained, because it was Aargau's natural resource agency that had set the stage for the project back in the 1990s. After noticing steep declines in amphibian numbers statewide, they had developed an action plan to address the problem.

"Build ponds," Schmidt summarized, giving an apt, two-word description for what grew into two decades of intense activity. He described the effort as "visionary," a rare collaboration among government agencies, politicians, conservation groups, and private landowners that nearly doubled the number of breeding sites in the canton. "It's really an obvious solution for amphibians," he said; more ponds should equal more habitat. But what wasn't so obvious, at least at first, was whether it would be enough to make a difference in one of Switzerland's most developed and densely populated regions.

"Aargau is not a pristine landscape," Schmidt said emphatically. "It is highly organized. There are roads, there are towns, there are farms using pesticides." For the project to succeed, frogs and newts would have to navigate such hazards, moving freely among old and new ponds to create what biologists call a "metapopulation." The concept works more or less like an insurance pool—a network of smaller populations that support and resupply one another as the need arises. "Some ponds always have breeding failure," Schmidt explained, pointing out any number of risk factors. They might dry up in a drought, or suffer a disease outbreak, or simply attract too many hungry predators. (As Schmidt put it, "Amphibians are food for lots of organisms.") In theory, adding new ponds to the system does more than make the metapopulation larger; it also makes it more stable, increasing the likelihood that there will always be productive sites to recolonize the ones that falter. In practice, theories about metapopulations often remain precisely that—theoretical—because it's simply too demanding to gather enough data for a field test. Computer models and genetic patterns offer insights, but few studies have actually tracked multiple subpopulations over long periods of time, particularly not for multiple species. The project in Aargau, however, was always intended to be accountable. Everyone from landowners to scientists to the bureaucrats providing the funding wanted to know if all their efforts (and investments) were paying off. And that meant one thing: counting a lot of amphibians.

"People like frogs," Schmidt mused. "Snakes or wasps would have been much harder." He went on to describe a monitoring program that had recruited and trained hundreds of volunteers to survey nearly every pond in Aargau at least three times per season on an annual or semiannual basis. It was an enormous task, adding up to tens of thousands of visits that, depending on the size of the pond, could last as long as an hour and a half. "Monitoring brings people together," Schmidt told me, but while the work certainly sounded challenging, he spoke of it with obvious fondness, as if the monitoring groups bonded less through adversity than through the sharing of a secret. Most amphibians are nocturnal, he explained, so the surveys often took place late at night when most people were indoors asleep, a situation that can make even densely settled landscapes feel empty and intimate. "It's dark, maybe it's raining a bit," Schmidt recalled, sounding almost wistful. "It's just you and that pond."

Over time, the scale of the volunteer effort in Aargau resulted in something more than data collection. "We had an army of amphibian friends on the ground," Schmidt explained, enthusiasts who helped to spread the word about the project and find new pond-building opportunities. "Most yards in Switzerland are very small," he reminded me, a fact that limited the number of landowners with space enough for even a few hundred square feet (less than one hundred square meters) of open water. Many ponds ended up in farm fields or in woodlands and parks on the edges of villages, but it soon became apparent that amphibians could find a pond almost anywhere. Again and again, surveys documented frogs, newts, and toads taking up residence at new sites, sometimes mere months after the ponds were dug. "We analyzed only presence or absence," Schmidt said, noting the difficulty of standardizing precise counts from so many different observers, on species that were sometimes hard to detect.[10] But even that simple metric turned out to be more than enough to see what was going on. After twenty years, population trends for ten out of the twelve species in the study had reversed completely, turning from decades of declines into

FIGURE 10.4. The construction of hundreds of small ponds like this one in Switzerland produced a cumulative region-wide impact, reversing population declines for ten different species of frogs, toads, and newts. Image © Benedikt Schmidt.

patterns of steady growth. A further species had stabilized, and only one, a toad that seemed to prefer larger bodies of water, continued to dwindle.

"Amphibian populations recovered!" Schmidt proclaimed toward the end of our call, and he emphasized that they did so in spite of all the challenges they faced in a highly developed landscape. Creating many small habitats had indeed created a positive cumulative impact, making Aargau one of few places on Earth where there are more frogs, toads, and newts today than there were thirty years ago. And, of course, amphibians are hardly the only creatures benefiting from all those new ponds. Schmidt mentioned how one of his graduate students was finding similar patterns of rapid colonization by dragonflies and aquatic snails, and he told me about another study that had noted positive impacts on swallows. "Their nestlings do better on a diet of aquatic insects," he said, which helps explain why adult swallows spend so much time swooping over open water during the breeding season, feasting on hatches of mayflies, caddisflies, and more.[11] A recent project in England echoed that result for a much broader array of birds, finding that pond restoration in farmlands boosted the diversity and abundance of at least two dozen avian species. Even ponds constructed for wastewater treatment have been found to harbor biodiversity, providing habitat for everything from native plants to insects, waterfowl, and crayfish.[12]

In Benedikt Schmidt's opinion, the results from Aargau should be seen as a call to action, a reason for all biologists to put a higher priority on restoration and recovery. "We spend too much time making lists of the threats to biodiversity," he said firmly. "We should focus more on solutions." That's not a bad way to think about any backyard biology project. It's true that there will always be species and habitats missing from our yards—they are not wilderness areas. But it's also true that there will always be room for improvement. And the solutions we find that boost plant and animal populations locally can add up to a much greater payoff at larger scales, like Schmidt's frog ponds, or Emily Geest's butterfly gardens, or Joe Buckler's vision for building up bird numbers, one custom nest box at a time. Yes, plants and animals face daunting threats on our warming, crowded planet, particularly in a society that seems increasingly disconnected from the natural world. But unlike most global challenges, this one can be addressed—in part, at least—right on our very doorsteps. When it comes to protecting biodiversity, the work begins at home.

CONCLUSION

Wild Crescendo

I come weary,
In search of an inn—
Ah! These wisteria flowers!

—Bashō (17th century)

Three sneezes left my head ringing like a struck bell, and I turned my face toward the sun, soaking up its welcome warmth. Studies show that moderate exposure to sunlight can reduce the symptoms of many respiratory infections by up to 15 percent, probably through a boost in the production of vitamin D. More to the point, sunshine feels good, a small compensation for the insult of catching cold in the summertime. With several work deadlines looming, I could ill afford taking a sick day, but my head felt too thick for writing, so I compromised by attempting a low-energy task that was at least work-related:

inspecting progress on what I hoped would become a buzzing, backyard bee garden. And if spending time in the sun helped me get well more quickly, all the better.

Wildlife managers always strive to provide what they call complete habitats, creating protected areas that encompass the food, water, shelter, and other resources their target species require to thrive. For pollinators, that means doing more than planting flowers—pollen and nectar can't sustain them year-round. Butterflies and moths, for example, also need specific host plants for their caterpillars to eat, and safe sites where they can form chrysalides and spin cocoons. Bees have similarly urgent requirements that go well beyond the flowers they feed on. To reproduce, they must have access to suitable locations and materials for building their nests. Mason bees and certain leafcutters will use holes drilled in wood, short lengths of bamboo, or a range of commercially available houses filled with cardboard tubes. Small carpenter bees are happy with old berry canes, pithy twigs, and other common garden waste. But an estimated 80 percent of wild bee species nest in the ground, and that's where things get tricky.

The best ground-nesting locations are sparsely vegetated, with soil conditions sandy enough for easy digging, but not so loose that the tunnels can't hold their shapes. Footpaths and the edges of driveways provide some backyard opportunities, but creating this kind of habitat on a larger scale is daunting. It involves scraping away the topsoil, replacing it with sand and small cobbles, and then using heavy machinery to compact the resulting surface. Few people would go to such lengths for the sake of bees. People who own horses, on the other hand, do it all the time, creating dedicated riding arenas with just the right footing for training and competition. If and when those arenas are abandoned, they become ripe for native bee restoration. The one at our place lies between my office and our vegetable garden and has been out of use for decades, a weedy legacy from previous landowners who built it as a sort of hub for local equestrians. My efforts to turn it into a hub for bees involved planting native flowers and trying to keep the other vegetation at bay, preserving as much open ground for nesting as possible. It's still very much a work in progress—the description "vacant lot" springs to mind. But every wild bee I see there counts as a small victory, and those victories add up.

From pollinators and potter wasps to brown creepers and breeding frogs, the creatures capable of thriving close to home are in many cases doing just that, aided measurably by the small steps we take to help them. So plant more flowers, put out a bird bath, and park that mower—it all makes a difference. Entomologist and biodiversity crusader Douglas Tallamy calculates that replacing half of the privately owned lawns in North America with native trees, shrubs, and wildflowers would collectively create more than five times the habitat found in Yellowstone and Yosemite National Parks combined. Over forty thousand people have already added their yards to this vision, joining a grassroots campaign that Tallamy calls Homegrown National Park. Dutch ecologist Carijn Beumer has coined an even catchier name for similar efforts to increase urban biodiversity in Europe: BIMBY, short for "biodiversity in my back yard."[1] But regardless of what you call it or where it happens, the growing interest in all forms of backyard biology (and the projects that measure it) is reconnecting millions of us with the natural world. It also provides a modest but much-needed counter to the litany of dire news we hear daily about our planet, from climate change to habitat loss, wildlife declines, plastic pollution, and more. In contrast to such global and seemingly intractable crises, the gains we make close to home can be counted bee by bee, blossom by blossom, and bird by bird.

All those elements were on display as I inspected the old horse arena, including a fuzzy-horned bumblebee nectaring on a native lupine, and two barn swallows flying in near tandem as they swooped and dove, hawking insects from midair. The scientist in me wanted to tally everything, to quantify just how much more wildlife was using the field since I began pulling weeds and scattering seed. But it struck me that no amount of data would answer a fundamental question that all of my work on backyard biology had yet to address: Why? Why should we bother protecting and restoring biodiversity—backyard or otherwise—in the first place? Debating such a basic point might sound gratuitous, particularly to people already passionate about the lives of plants and animals. (It certainly never came up during the interviews I conducted for this book.) But convictions alone, no matter how deeply held, can't convince a skeptical neighbor, and they rarely budge the needle on policy. Clear, relatable answers to the "Why?" question can help advance the cause of biodiversity conservation in any setting. While textbooks offer a range of theories and justifications, I've always found that the main themes can be summarized with anecdotes from three improbable but surprisingly suitable sources: Spider-Man, Scrooge McDuck, and Mickey Mouse.

The Spider-Man origin story, told and retold through various print and film adaptations for over half a century, features young Peter Parker struggling to adapt to his role as a superhero. Bitten by a radioactive spider, he develops strange new abilities and must decide how to use them, giving rise to the overriding theme of the series: with great power comes great responsibility. That dictum is a good ethical signpost for any aspiring superhero, but it has meaning in the real world too. How better to describe the quandary of life on a human-dominated planet? Never before have the actions of a single species determined the fates of so many others, a power imbalance that, by Spider-Man's way of thinking, comes with unavoidable obligations. The argument is one of stewardship, the idea that humanity's great influence over natural systems gives us a caretaker's duty to safeguard them, and to protect the species that call them home. It applies at any scale but seems particularly well suited to backyards and neighborhoods, the settings where people have the most direct control over how land is managed. The reasoning behind this approach is indisputable—human decisions and activities will certainly determine the future of biodiversity on this planet. But, at its heart, it is a moral argument. It relies on the notion that all species have intrinsic value, and that protecting them is the "right" thing to do. That may be true, but following through on moral arguments ultimately relies on goodwill, and even people who like plants and animals are often unwilling to make major efforts or sacrifices on their behalf. For those situations, other justifications are required, which leads us straight to the philosophy of a very different comic book icon.

Fans of the early Disney universe will remember Donald Duck's fabulously wealthy and famously tightfisted uncle, Scrooge McDuck. Known for hoarding his fortune in a giant Money Bin on a hill above Duckburg, and occasionally swimming around in its deep piles of cash, Uncle Scrooge is portrayed as a shrewd business-duck who never does anything without a profit motive. Where Spider-Man is preoccupied with responsibility, Scrooge is cartoonishly selfish, living by a credo that can be neatly summed up in five words: "What's in it for me?" This attitude often appears in the real world too, where it can provide another strong argument for protecting biodiversity. That's because regardless of whether someone believes that other species have value intrinsically, no one can deny that many of them are extremely valuable to people.

Nature-based industries from farming to fisheries to forestry all depend utterly on other species, and not just the domestic or wild-harvested ones. There is always a diverse cast of supporting biological characters making such activities possible (and profitable). Global pollination services, to cite one well-studied example, have been valued at close to $400 billion annually. But that number is in some ways beside the point, because no one actually believes that any amount of spending could replace bees and other pollinators. The cost of hand-pollination only pencils out for specialty crops like vanilla beans and dates, and recent attempts to engineer pollinating drones have all been comically clumsy. Unsung species underpin other business models too, from the wild marine food webs that fatten up commercial fishes to the soil fungi fueling the growth of fir trees. In the medical trade, bioprospecting has provided the botanical basis for many familiar and highly lucrative drugs, including aspirin (from willow bark), Coumadin (from tonka beans), Taxol (from yew trees), and the many derivatives of morphine (from poppy latex). Other cures also involve surprising elements of biodiversity—development of the Novavax coronavirus vaccine, for example, involved growing spike proteins inside the cells of an owlet moth caterpillar and then adding in a dash of extract from the Chilean soapbark tree.[2] Innovations of all kinds often take their cues from other species. Biomimicry has given us everything from wood-based paper products (inspired by the nest-building habits of wasps) to Velcro (inspired by the way burdock burrs tangle in dog fur) to a new generation of sticky surgical adhesives (inspired by the slime exuded from slugs). In a backyard context, the simplest dollar-value calculation for biodiversity can often be found in the price tag of the habitat itself. Since the 1970s, studies of real estate transactions have consistently found that the presence of mature trees and shrubs increases property values, sometimes by more than 10 percent.[3] Premiums are also paid for other backyard vegetation—not just lawns, but flower beds, hedges, and even rockeries dotted with cacti and succulents. In short, Uncle Scrooge could find any number of ways to make money from biodiversity. But there is also a third line of reasoning to be considered—one that has less to do with value than it does with risk. Unexpected consequences often crop up when we mess around with things that we don't understand, a theme perfectly captured in the most famous role ever played by the world's most famous rodent.

As the title character in *The Sorcerer's Apprentice*, Mickey Mouse gets himself into serious trouble when he dons his master's magic hat without permission and commands a broom to do his chores for him. Soon all hell breaks loose when the enthusiastic (and increasingly disobedient) broom overflows a basin with buckets of well water, flooding the workshop. By the time the old sorcerer returns, one broom has multiplied into an army and Mickey is awash, paging desperately through the spell book in search of a fix. We run similar risks when we lose biodiversity because the natural world is very much like a spell book we have yet to master. Complex relationships link seemingly disparate species, making it impossible to predict the consequences of removing particular plants or animals from the mix. Insights often come only after species are long gone, or, in a few cases, after they've been returned.

When gray wolves were reintroduced to Yellowstone National Park following an absence of seventy years, everyone was pretty sure they would impact the local elk population. Fewer people predicted their impact on gooseberries. Or trout. Or flycatchers, beavers, song sparrows, alder trees, willows, and scores of other species now caught up in the story.[4] It turns out that wolves meant more to the ecosystem than a single predator-prey interaction. Their return not only reduced the density of elk herds, it kept them on the move, allowing the shrubs, trees, and other vegetation along rivers and streams to recover from decades of overbrowsing. That new growth transformed riparian woodlands, as well as the banks and channels of streams, with implications for all the species that call those habitats home. Similar patterns occur among reintroduced wild dogs, waterbuck, and floodplain vegetation in Mozambique's Gorongosa National Park, while the return of fin whales to their historical feeding grounds in Antarctica has reinvigorated what experts call the "whale pump," a cycle of nutrients transported from the depths (where whales feed) to the surface (where they defecate), fertilizing plankton growth and impacting the entire food web. Biologists may argue about the scale of such effects, but few contest that the removal (or addition) of species can trigger widespread change.[5] Nor do the creatures involved need to be large and showy. Unpredictable, community-altering consequences have followed experimental manipulations of everything from starfish on intertidal rocks (the original keystone species), to crawdads in boreal ponds, to beetles and ants in tropical forests. Perhaps the boldest application of these ideas, however, occurs in paleontology, where the mass extinction of Pleistocene megafauna (e.g., woolly mammoths, saber-toothed cats, giant kangaroos) has been suggested as a driver of continent-scale environmental transformations. By fundamentally restructuring patterns of predation and herbivory, the rapid loss of so many species may have flipped the arctic from steppe grasslands to tundra, and changed Australia's dominant vegetation from a rainforest to an arid, fire-prone scrub.

In many cases, biodiversity is also thought to increase the productivity and stability of ecosystems. Evidence for this comes in part from a famous study that sounds like precisely the kind of project that would happen in someone's backyard. Researchers planted mixtures of grasses, flowers, and shrubs in 30-by-30-foot (9-by-9-meter) plots and mowed portions of them annually for ten years, sorting and weighing the clippings. In every instance, plots with more species produced more foliage, and they did so more consistently over time. Scaling up and adding animals to the mix seems to amplify that pattern. In a recent global analysis, climate scientists found that ecosystems with intact species complexity, where plants were joined by seed dispersers, grazers, predators, and all the other players in a functioning food web, stored from 15 to 250 percent more carbon than those with pieces missing.[6] No one knows precisely why. Faced with the complexity of natural systems, even the most experienced biologists often feel as flummoxed as apprentices in ill-begotten hats. But unlike the Mickey Mouse cartoon, one thing in the real world is certain: no master sorcerer will come back at the end of the story to set things right. If we allow biodiversity to dwindle, we'll simply have to live with the consequences.

There is another benefit derived from biodiversity that appears to be universal but that often goes unrecognized because it boils down to something intangible: how nature makes us feel. Instead of a comic book analogy, the best illustration of this principle has to do with hospital beds. In 1984, a two-page report appeared in the journal *Science* describing a simple and surprising study from a small suburban hospital in the state of Pennsylvania.[7] The authors analyzed ten years of data on patients recovering from the same gall bladder surgery in the same ward. After controlling for such variables as the age, sex, and general health of the patients, only one thing separated the cases under consideration. Some of the patients were lying in beds where the view from the window featured a grove of trees. The others looked out at a brick wall. In case after case, the results were the same. Patients with a view of trees recovered more quickly, required fewer painkillers, made fewer calls to the nurses' station, and experienced fewer complications from surgery. To be clear, the trees in question were not towering, old-growth behemoths, breathtaking in their majesty—it wasn't necessary to be awed to heal faster. In a conclusion that should resonate with any backyard biologist, the mere presence of typical suburban trees, viewed through glass, was enough to enhance the process of healing.

In the decades since its publication, the hospital bed study has become the research equivalent of a runaway bestseller, cited by more than eight thousand other papers that bear out and expand upon its findings. Confirmed positive effects of nature on human health range from stress reduction to lowered blood pressure, reduced cancer mortality, lower incidence of diabetes, and, for pregnant mothers, higher infant birth weights.[8] The evidence is so overpowering it is now possible to receive a medical prescription for nature exposure from doctors in thirty-five states, four Canadian provinces, and dozens of countries around the world, including Japan, where the remedy is charmingly referred to as forest bathing. Savvy practitioners will indeed specify a forest or other rich habitat, since green spaces with higher biodiversity often produce better results, particularly when that diversity includes easily recognized groups like birds and butterflies.[9] In other words, people feel even better when they know they are surrounded by a rich mix of species—all the more reason to encourage biodiversity in the places close to home.

The effects of fledgling bee gardens on human health have yet to be studied directly, but I still viewed my time inspecting the old horse arena as a form of self-medication. Truth be told, my head cold wasn't the only thing making the day harder to embrace than most. Other life stresses were involved, and, as the saying goes among biologists, "Nature is cheaper than therapy." I found a patch of gravelly soil warm from the sun, and lowered myself down to sit amongst a scattering of wildflowers—the pink topknots of sea blush, the vibrant stars of blue-eyed grass, and the rosy, four-petaled faces of farewell-to-spring. There was a faint breeze making the blossoms nod, as if tugged by invisible strings, and I noticed the leaves of a gumweed I'd planted and forgotten all about—small, but still surviving in the thin, dry soil. Then a bee appeared, a jet-black miner with a wash of tawny fuzz like a velvet stole across its thorax. I watched as it hovered and drifted back and forth, inches above the cobbles and vegetation. It was a species that I'd never seen before on our property, a ground nester more common in the remnants of native prairie that dot our island's shoreline, several miles distant. Would it nest here? I held my breath when the bee landed, but then lost sight of it as it crawled off amongst the tufts of grass and wildflowers. My mind drifted, and a few minutes later I decided to heave myself up and head back to the office for another try at indoor work. Just then, a furtive motion caught my eye near the base of a bunchgrass. Turning, I saw it again, and felt a rise in spirits beyond the measures of science. It was sand, a flicker of grains flung skyward by tiny legs. The bee had begun to dig.

Acknowledgments

Ushering books into the world is a process that involves many hands. I am grateful, as ever, for the guidance of my wonderful agent, Laura Blake Peterson, and so pleased to be working again with editor Thomas Kelleher, publisher Lara Heimert, and all their superlative colleagues at Basic Books, including Madeline Lee, Kristen Kim, Gillian Sutliff, Amber Hoover, Laura Piasio, Melissa Veronesi, Susie Pitzen, and no doubt many others behind the scenes. My thanks extend to the many booksellers and librarians who find room for my work on their shelves, and of course to the people who choose to invest their precious reading time in my efforts—without them, the gears of this whole endeavor would quickly grind to a halt. On the home front, Eliza and Noah remain foundational to everything that I do, showing infinite patience with Papa's many odd habits, from sitting in trees and stick houses to taking up an inordinate amount of freezer space with jars of specimens. I am also fortunate to enjoy the support of my extended family, many friends, and a community of like-minded souls on the island that I call home.

Finally, let me extend heartfelt thanks to the biologists and fellow backyard biology enthusiasts who have spoken with me about their work and experiences. Any mistakes in relating their research and stories in these pages are mine alone. Here, in no particular order, are some of the many people who have contributed, collaborated, and otherwise helped this project along the way: Emily Hartop, Christine Gabriele, Phil Green, Michelle Fournet, Alan Irwin, Boyd Pratt, Doug McCutcheon, Lisa Gonzalez, Alexander Fateryga, Kiri Ross-Jones, Douglas Tallamy, Gretchen LeBuhn, Kolla Sreedevi, Brian Brown, Daniel Martin, Dirk Steinke, Scott Loarie, Charles Abramson, Barbara Klump, Steve Sillett, Peter Neitlich, Anne Goodenough, Charley Eiseman, Julia Blyth, Jacqueline Beggs, David Nowak, Tawm Perkowski, James Carpenter, Sophie Lokatis, Joe Buckler, Philip Matthews, Emily Geest, Emma Webb, Candice Rennels, Benedikt Schmidt, Mark Anthony, Richard Bardgett, Robert Cichewicz, Jeffrey Ethier, Christopher Kyba, Kevin Gaston, and Sibylle Schroer.

Appendix

Citizen Science Resources

The practice of backyard biology can be as simple as pausing to watch a bee on a flower, or deciding to leave a patch of grass unmowed. But there are also countless opportunities to participate in more organized efforts, loosely grouped under the umbrella of citizen science. Many projects have been mentioned in these pages, and those are listed alphabetically below, alongside their web addresses. As a first step, however, I highly recommend perusing one of the major citizen science portals, such as SciStarter (www.scistarter.org) or Zooniverse (www.zooniverse.org). They catalog thousands of projects and allow searches by topic, helping anyone track down subjects of particular interest, from ladybug surveys in London, to the sounds made by geckos in Singapore, to iSeahorse, an effort to document seahorse populations on coastlines around the world.

Antibiotics Unearthed: www.microbiologysociety.org
BioSCAN: www.nhm.org/community-science-nhm/bioscan
BugGuide: www.bugguide.net

DarkSky International: www.darksky.org
Drugs from Dirt: www.drugsfromdirt.org
eBird: www.ebird.org
Globe at Night: www.globeatnight.org
GLOBE Observer: https://observer.globe.gov
The Great Sunflower Project: www.greatsunflower.org
Homegrown National Park: www.homegrownnational park.org
iNaturalist: www.inaturalist.org
i-Tree: www.itreetools.org
KraMobil: www.spotteron.com/kramobil/info
National Moth Week: www.nationalmothweek.org
Taxon Expeditions: www.taxonexpeditions.com
TreeSnap: www.treesnap.org
Treezilla: www.treezilla.org
University of Oklahoma Citizen Science Soil Collection Program: on hiatus; data archived at www.shareok.org/handle/11244/28096

Notes

Introduction: Backyard Biology

1. Literary references to the hermit thrush and its remarkable voice probably would have started earlier, but the birds are so shy and secretive that ornithologists didn't observe them singing until the mid-nineteenth century. See Doolittle 2020.

2. Whitman 1976, p. 267.

3. Experts define daily interactions with nature broadly, from hiking in a national park to playing in a vacant lot. The strength of the correlation varies by activity, but the link between time spent outdoors and concern for the environment is remarkably consistent. See Rosa and Collado 2019 for a review, and see Frumkin et al. 2017 for a discussion of an added bonus: health benefits. Simply getting outside also contributes to stress reduction, improved sleep, lower blood pressure, reduced diabetes, better eyesight, and more.

4. The vast majority of anthropause observations were anecdotal, including the barn swallows at our local bank that nested above what had previously been a busy ATM machine. But in some cases, biologists were in a position to take accurate measurements. Good examples include Driessen 2021 (for the drop in wallaby roadkill) and Derryberry et al. 2020 (for changes in sparrow song). See Bates et al. 2021 for a thoughtful review.

5. There is a wide literature on the remarkable diversity of plants and animals used by hunter-gatherers. African examples cited in this discussion include Lee 1978 and Terashima and Ichikawa 2003, but also see Schultes and Raffauf 1990 for an exhaustive botanical catalog from South America, and Moore et al. 2000 for an intriguing paleolithic example unearthed at excavations in Syria.

6. In addition to the natural history observations sprinkled throughout his writing, Thoreau kept meticulous notes on the timing of seasonal events like bird migrations and nesting periods, wildflower blooms, and leaf senescence. His data provide a rare and important baseline for gauging modern shifts associated with global warming. As botanist and modern Walden Pond scholar Richard Primack once told me, if Thoreau were listed as a coauthor on all of the papers now using his data, he would be one of the most prolific climate change scientists of the twenty-first century!

Chapter One: Look at Your Fish

1. Humble, wooden structures can sometimes seem like a job requirement for nature writing. Henry David Thoreau kick-started the trend with his "shanty board" cabin at Walden Pond (Thoreau 1966, p. 33), and many others have followed suit. Aldo Leopold scribbled away in a converted chicken coop, for example, and the walls of John Burroughs's writing retreat were so rough he called it "Slabsides." It took the likes of Pulitzer Prize winner Annie Dillard to upend this tradition, evoking the charms of Tinker Creek from a study room at her local library, with cinderblock walls and a single window overlooking a parking lot.

2. Anonymous 1874, p. 369. Samuel Scudder published his remembrance of Louis Agassiz anonymously, a common practice at the time. The essay first appeared in the popular magazine *Every Saturday* in 1874 and was reprinted widely thereafter under the title "In the Laboratory with Agassiz," attributed simply to "A Former Pupil."

3. Ibid., p. 370.

4. Ibid.

5. Ibid.

6. Ibid.

7. Ibid.

8. Charles Darwin's flair for a creative experiment arose long before he settled in at Down House. As a schoolboy, his penchant for homespun chemistry earned him the nickname "Gas." And in one memorable episode during his voyage on the *Beagle*, he hid a piece of meat in white paper and presented it to several captive Andean

condors in an attempt to gauge their sense of smell. "I walked backwards and forwards, carrying it in my hand at the distance of about three yards from them," he wrote, "but no notice whatever was taken of it" (Darwin 1838, pp. 5–6). For a thorough and entertaining exploration of Darwin's experimental habits, see Costa 2017.

9. Nash 1890, p. 406.

10. Published as a follow-up to *Origin of Species*, Darwin's orchid book appeared in 1862 under the title *On the Various Contrivances by Which British and Foreign Orchids Are Fertilised by Insects, and On the Good Effects of Intercrossing*. Sales were disappointing. Nonetheless, it contained groundbreaking observations and ideas at a time when the process of pollination was still poorly understood, and the role of insects remained controversial.

11. Habitat maps paint the coastal forests of the Pacific Northwest with a broad brush, but conditions and tree species vary considerably depending on latitude, local topography, and elevation. Although the landscape is densely wooded with conifers, only the wettest places receive the 55 inches (140 centimeters) or more of annual precipitation necessary for strict "rainforest" status. Our island woods receive far less, lying partially within a rainshadow created by the nearby Olympic Mountains.

12. Darwin 2004, p. 446.

13. From a letter to W. J. Hooker dated April 9, 1833, as quoted in Harvey 1947 (p. 196) and confirmed via correspondence with Kiri Ross-Jones, archivist, Kew Gardens.

14. Mayne 1862, p. 89.

15. From a letter to J. D. Hooker dated June 5, 1855. See Darwin Correspondence Project 1855.

16. Appearing in various forms and translations, this saying can be traced to a longer passage from Proust's *Remembrance of Things Past*: "The only true voyage of discovery, the only fountain of Eternal Youth, would be not to visit strange lands but to possess other eyes, to behold the universe through the eyes of another, of a hundred others, to behold the hundred universes that each of them beholds, that each of them is." Proust 2006, p. 657.

Chapter Two: Get Small

1. There is a rich scientific literature devoted to attention restoration theory, the idea that time in nature improves mental health by (in part) offering "softly fascinating" stimuli that encourage "indirect attention." Though not without controversy, the theory rests on the notion that we have an intrinsic relationship to the natural environment, a set of evolved responses that echo our history as a species. See Kaplan 1995 for the classic paper, and Stevenson et al. 2018 for one of many recent reviews.

2. Wallace 1872, p. 222, 422.

3. Darwin 1908, p. 196.

4. Published in 1859, the same year as *Origin of Species*, this less heralded Darwinian contribution in *The Entomologist's Weekly Intelligencer* announced three rare beetles "taken lately, in the parish of Down, six miles from Bromley, Kent." Though certainly written and submitted by Charles, it was signed "Francis, Leonard and Horace Darwin" (ages ten, eight, and seven, respectively). See Darwin et al. 1859.

5. Fabre 1914, p. 7.

6. Ibid., p. 8.

7. Ibid., p. 8.

8. Common throughout the Mediterranean region, the black mud bee or mortar bee is known by the Latin binomial *Megachile parietina* (*Chalicodoma muraria* in Fabre's era). In the wild it constructs its nests on the sides of pebbles and rocks, but in towns and cities it has adapted to all sorts of hard surfaces, including bricks and blocks of stone, giving rise to another common name: the wall bee.

9. Though he rejected Darwinian evolution and made few contributions to theory or taxonomy, Jean-Henri Fabre remains esteemed as the quintessential observer, describer, and chronicler of insect lives. His many books earned him two nominations for the Nobel Prize, and reportedly led Victor Hugo to dub him "the Homer of insects." For more on his remarkable life and work, see Farvet 1999.

10. Published by the Smithsonian Institution, this venerable reference has not been updated since 1979. Given the advances in

taxonomy in the decades since, as well as the many varieties yet to be named, the real number of wasps in America north of Mexico is undoubtedly much higher. See Krombein et al. 1979.

11. Fabre 1919, p. 5.

12. Swift 2010, p. 43.

13. Known to botanists as "glandular trichomes," these sticky hairs appear on as many as 30 percent of all plant species. They present attacking insects with a physical hindrance, but also contain a huge array of defensive chemicals, including terpenes, phenols, and volatile oils. Familiar aromatic herbs from marjoram to mint, basil, sage, and oregano all rely on trichomes to produce their distinctive flavors, and the booming marijuana and medicinal hemp industries can thank trichomes for the active ingredients in cannabis and CBD oil.

14. Desirée Narango, one of Doug Tallamy's graduate students and now a postdoctoral fellow at the University of Massachusetts Amherst, studied the relationship between native plants, caterpillars, and Carolina chickadees across ninety-seven yards in Washington, DC. The startling estimate of caterpillar consumption stems from her work and from basic natural history observations compiled in the 1950s. See Narango et al. 2017, 2018 and Brewer 1961.

15. Doug Tallamy told me he always tries to avoid using the word "backyard" when describing his habitat restoration activities. "It implies something we need to hide away," he explained, "but that's not true. You can plant that native oak tree in your front yard too!"

16. Precisely what attracts moths and other insects to artificial lighting remains unclear, but it appears to simply overwhelm their navigational systems. Many nocturnal insects orient themselves, at least in part, by subtleties of moonlight and starlight. The bright, sustained, ground-level illumination introduced by human technologies has no equivalent in nature. For insects, it isn't just unfamiliar; it is outside of their evolutionary history. See Owens and Lewis 2018.

Chapter Three: Something New

1. When I contacted Brian Brown he made himself immediately available and spoke with the passion and modesty of a true

enthusiast. "You can see I'm reluctant to talk about this stuff," he joked at one point, nearly an hour into our conversation. Later, I combed through the literature and did a back-of-the-envelope calculation to better understand his outsized role in the world of phorid flies. Roughly 4,500 species have been identified to date—Brown has played a role in discovering and naming more than 700 of them.

2. The head usually drops from the ant's body as the phorid larva nears maturity. This allows the young fly to finish its development and pupation on the ground, safely enclosed in the hard shell of the ant head. Among scores of ant-decapitating species, several phorids in the genus *Pseudacteon* specialize in attacking South American fire ants, a destructive invasive species in parts of Australia, Asia, and the southern United States. Ongoing research points to phorids as a potentially game-changing method for controlling fire ant outbreaks. See Chen and Porter 2020.

3. Wilson 1987, p. 344.

4. Combining modern genetic tools with traditional methods has dramatically boosted the pace of species discovery, particularly for insects, fungi, bacteria, and other cryptic groups. Scientists now describe roughly twenty thousand new species and varieties every year, leading some to call the twenty-first century "the new age of discovery." But even at that brisk rate, it will take over four hundred years to put names on all of the Earth's estimated eight million unknown life-forms. In other words, there is good job security in taxonomy! See Grieneisen et al. 2014.

5. The species discoveries mentioned in this paragraph include Fontaine et al. 2012 (Europe 1998–2007); Solomon 2022 (mushrooms); Oh et al. 2019 (jewel anemones); Martin et al. 2022 (bristle worms); Korshunova et al. 2021 (nudibranch); van Achterberg et al. 2020 (wasp); and Schilthuizen et al. 2017 (taxonomy-themed ecotours).

6. Mentored by taxonomists at the University of Adelaide, the students in South Australia targeted specimens from only one subfamily of small wasps. Using a broader approach (and a lot more schools), the School Malaise Trap Program in Canada identified 1,288 specimens whose genetic barcodes didn't match anything on

file. Most of those were probably varieties that hadn't yet been studied genetically, or that had never been recorded in Canada, but the scientists behind the project expect further analysis to almost certainly reveal many new species. See Steinke et al. 2017 and Fagan-Jeffries and Austin 2021.

7. The original iNaturalist team consisted of Ken-ichi Ueda, Nate Agrin, and Jessica Kline, all graduate students at the University of California, Berkeley. The others soon moved on, but Ueda remains as cofounder and codirector, keeping the platform rooted to its original goal as a place to, in his words, tell other naturalists, "I saw this cool thing!" (Ueda 2020).

8. Ueda 2020.

9. Since I spoke with Scott Loarie in late 2022, those figures have continued to rise. As this book goes to press, iNaturalist users have posted a staggering two hundred million observations of nearly 450,000 different species.

10. Ken-ichi Ueda and Scott Loarie ran iNaturalist on their own for several years before picking up important sponsorships from the National Geographic Society and the California Academy of Sciences. As a nonprofit organization with a staff of just eleven full-time employees, iNaturalist still relies heavily on donations and grants, and the dedicated services of legions of online volunteers who help curate and monitor the site.

11. GBIF 2022.

12. Although an invaluable resource for general inquiry, GBIF was never envisioned as a replacement for natural history collections. Examining physical specimens remains vital for many research questions about evolution and biogeography, including genetic analyses and comparisons of fine anatomical details not visible in digital photographs.

13. The variety of studies citing data from iNaturalist (and other citizen science sites) is truly stunning, constantly expanding, and worthy of perusal. Just visit a science database site like GBIF or Google Scholar and use the search term "iNaturalist." Studies mentioned in this paragraph include Goldstein et al. 2018 (dune grasses); Cheng et al. 2019 (Hong Kong moths); Tanaka et al. 2021 (juvenile

great white sharks); Shin et al. 2022 (Korean roadkill); Mo and Mo 2022 (iguanas); and Moore et al. 2021 (dragonfly wing spots).

14. Jain et al. 2022.

15. The specimens in question turned up when Washington State entomologists collected the first Asian giant hornet (*Vespa mandarinia*) nest ever found in the United States. It's a testament to the great adaptability of phorids that they had already learned to scavenge inside the home of such a recent arrival, an alarming insect that most other local species seemed happy to avoid. (They're not called "murder hornets" for nothing.) It's also fitting, from the standpoint of backyard biology, that after months of searching, including the use of hornet-sized radio tracking devices, officials finally located the nest in a hollow tree in someone's yard, less than 100 feet from their house, sharing space with a swing set and dog kennel.

Chapter Four: Forces of Habit

1. From "The First Robin" by Albert Laighton (Laighton 1878, p. 66).

2. Kimball 1944, p. 646.

3. Ibid.

4. I once witnessed a striking example of this principle while studying scavenger birds in East Africa. Marabou storks dwarfed the other species attracted to carrion, but their massive, pointy bills were useless for opening carcasses. That task fell to the hooked bills of vultures, hawks, and eagles. As soon as the skin was breached, storks moved in and dominated the struggle over entrails and other soft tissues. But their reign was brief. When it came time to tear meat from bone, the storks once again lacked the right equipment and were forced to hand the carcass back to their hook-billed competitors.

5. See Partners in Flight 2022.

6. The woodpecker in question was a northern flicker, a species that typically feeds on the ground. So it already had my attention when it began tapping high up in a tree. I was as startled as the owl when the branch dropped. The behavior was almost certainly intentional—smaller songbirds had already been mobbing the owl,

and the flicker had no other conceivable reason to land above them at that moment in that particular tree. If it did indeed use a branch to move an owl, then the flicker came remarkably close to meeting the common definition for animal tool use: "the external employment of an unattached environmental object to alter . . . the form, position, or condition of another object, another organism, or the user itself" (Beck 1980, p. 10).

7. Turner 1908, p. 253.

8. Ibid.

9. Ibid.

10. Ibid., p. 257.

11. Some biographers have suggested that Charles Henry Turner eschewed academia, preferring instead to teach at the high school level. But he is known to have held positions briefly at the University of Cincinnati and at Clark College in Atlanta, and was turned down for at least one position at the University of Chicago. It's unknown how many other research jobs he applied for over the years, but it's hard to imagine that racism didn't severely limit his career options. See Abramson 2006, 2009.

12. Du Bois 1939, p. 309.

13. Ibid.

14. The existence of animal culture remains a hotly debated subject in biology. Like tool use, it was long considered a distinctly human trait until evidence from other species began to pile up. In addition to examples from primates, whales, and birds, putative cultural traditions (or at least examples of social learning) have been observed among populations of everything from bighorn sheep to rats, sticklebacks, and bumblebees. See Whiten 2021 for a review, and Laland and Janik 2006 for a critique.

15. Primatologists witnessed a similar behavioral origin story in the 1950s on the island of Koshima in Japan, where a single macaque developed the habit of washing sweet potatoes in a stream before eating them. Other macaques copied her and the behavior soon spread through the entire population (see Kawai 1965). Klump's cockatoo study follows suit, but also shows how remarkably quickly

an innovation can diversify, becoming culturally distinct among subpopulations after only a year or two of partial isolation.

16. When I asked Barbara Klump if citizen scientists would play a role in her crow study, the answer was an adamant "Yes!" She has already begun amassing reports on nut-dropping crows via the cell phone app KraMobil. The project is a major theme in her new lab at the University of Vienna, where she recently accepted a post as assistant professor of cognitive sciences.

17. As insights from her work continue to pile up, it's not surprising that Klump is also giving thought to a very basic, theoretical question: Why do animals develop culture in the first place? Culture is an inherently social phenomenon, she explained, "but not every social species has culture." Those that do tend to share certain traits like large brains, high intelligence, and a long period of juvenile development. Sulphur-crested cockatoos, for example, spend seven years as juveniles. But Klump shared with me another item for that list that I'd never heard before. "Long-lived species have less chance to adapt through evolution," she told me, and pointed out that creatures like cockatoos, chimpanzees, and whales lived for decades, producing fewer offspring, over longer periods of time, than most other organisms. It was a matter of simple mathematics. Less frequent breeding events meant that long-lived creatures didn't "have the same opportunity to adapt through genes." That situation favored the development of culture as a sort of evolutionary workaround, she argued. It gave the animals a way to quickly master new behaviors to help them adapt during their lifetimes rather than over the course of generations.

18. Millions of starlings from northern Europe migrate south to winter in places like Rome, where huge murmurations sometimes appear above the Colosseum and other major tourist attractions. Their droppings, often greasy from feeding on olives in the surrounding countryside, can become thick enough to cause traffic accidents, close roads, and damage property. The city's environmental department maintains teams of dedicated starling-chasers trained to drive the flocks away from problematic roost sites, blasting them with prerecorded starling distress calls on loudspeakers and bullhorns.

19. Behavioral adaptations abound in urban settings (see Lowry et al. 2013), and evidence for urban-driven evolution is compelling, if often incomplete (see Lambert et al. 2021). Key studies mentioned here include Bolliger et al. 2020 (bats at streetlights); Debrot 2014 (birds at streetlights); Brown and Brown 2013 (swallows); Harris and Munshi-South 2017 (mice); and Ossola et al. 2018 (publication trends).

Chapter Five: Above

1. Although first published in 1882, this favorite Nietzsche aphorism did not appear in English until the twentieth century. One early translator used the phrase "unexplored seas" (Nietzsche 1910), but the more commonly quoted version appears as "uncharted seas" (e.g., Archer 1915, Nietzsche 1974).

2. My research as a doctoral candidate involved tracking the pollen dispersal of large rainforest trees. Curious about the insects doing the pollinating, I briefly entertained the notion of climbing into the canopies of Central American jungles, crawling out to the tips of branches, roping myself near clusters of flowers, and swinging a net at passing bees. Taking a tree-climbing course, where I failed to even get myself safely aloft in a manicured oak on the university quad, quickly brought me to my senses.

3. Erwin 1982, p. 75.

4. Erwin 1983, p. 14.

5. Humbly published in *The Coleopterists Bulletin* in 1982, Erwin's two-page beetle paper has become foundational to our understanding of the magnitude and diversity of life on Earth, taught in college classrooms for generations, reproduced in textbooks and compilations, and cited by other researchers more than 1,500 times. Erwin went on to refine and repeat his studies in other locations, particularly the Amazon Basin, where he found even higher proportions of beetles specific to particular kinds of trees. Samples collected only 150 feet (50 meters) apart under the same forest canopy in Peru, for example, differed in their makeup by more than 90 percent, leading Erwin to up his estimate of the Earth's insect diversity to more than 50 million species. That number remains controversial, and the true

total is unknowable, but Erwin's work provides a striking glimpse of the mind-boggling diversity found in tropical forests. It's also a sobering reminder of what is being lost as they continue to be cleared. See Erwin and Scott 1980, and Erwin 1982, 1988.

6. The lichen in question belongs to one of the world's most vibrant and beautiful genera, *Chrysothrix*. It's considered a crust-forming, or crustose, species (in contrast to leaf-like foliose or branching fruiticose lichens) and consists almost entirely of tiny bright yellow fragments called soredia. Each of these little motes measures less than five thousandths of an inch across (0.1 millimeters) and contains bits of the lichen's distinctive combination of fungi and algae, ready to disperse and replicate in a new location. The vibrant yellow color comes from pigments like pinastric acid or vulpinic acid that are thought to help regulate internal temperature and shield the lichen's resident algae from overexposure to solar radiation. See Brodo et al. 2001 for an authoritative discussion.

7. Look closely and you will find a flaky patina of lichens covering (and coloring) the bark of virtually any tree, not to mention other familiar outdoor surfaces including rocks, sidewalks, deck railings, and, at our house at least, the roofs of unwashed cars.

8. Ancient forest lichen communities aren't just diverse, they are abundant. In one study, the lichen biomass adorning a 510-year-old forest canopy weighed thirty times as much as the scant covering found in a 40-year-old canopy nearby, and triple that of a mature, 140-year-old stand. That lush growth provides unique habitat for arboreal insects, birds, and animals, but it also contributes a surprising boost to the fertility of forest soils. Nitrogen fixed from the air by lichens reaches the forest floor when they fall and decompose, providing up to 50 percent of all nitrogen inputs in some old-growth systems. See Denison 1973, Neitlich 1993, and Preston 2007.

9. Even if the Douglas fir in our yard reached three times its current height, it would fall well short of the record holder for the species, a 393-foot (120-meter) giant that toppled in a windstorm near the town of Mineral, Washington, in 1930. Anecdotal evidence from early European settlers and loggers points to numerous specimens well in excess of 400 feet (122 meters).

10. Similar in appearance and habit, several species from the genus *Dolichovespula* are also commonly referred to as yellowjackets. Taxonomists know them as the long-faced (*dolicho-*) *vespula*, based on certain cheek measurements. For everyone else, they can be thought of as aerial yellowjackets, a reference to their taste in real estate. Aerial yellowjackets build football-sized paper nests in high, open places, dangling from the branches of trees and shrubs, or the eaves of buildings. Yellowjackets in the genus *Vespula* also build paper nests, but you never see them. They prefer locations hidden away underground, or in tree cavities, hollow stumps, and gaps in the walls of sheds and houses.

11. Bostock and Riley 1855, p. 24.

12. Holland 1601, p. 322.

13. As a general rule, it takes protein to build protein, and wasps are far from the only creatures to eat more of it during periods of active growth. Larval honeybees feed on protein-rich pollen, for example, before switching to nectar and honey as adults. For songbirds, a diet heavy on caterpillars and other invertebrates is common for nestlings, even in species that focus on seeds and fruit later in life. Hummingbirds also seek out more insects during the breeding season, for themselves as well as for their offspring, sometimes forgoing nectar entirely for weeks at a time. See Montgomerie and Redsell 1980.

14. A notable exception to the xylem rule occurs in early springtime, when certain trees move sugars stored in their roots upward through the trunk to fuel new growth in their shoots and leaves. Tapping maple trees during this period provides the sap used to make syrup, but in its raw form no one would want to put it on a stack of pancakes. Fresh xylem sap from a maple tree is thin and watery, with a sugar concentration of only 2 or 3 percent. The transformation into syrup takes place inside a sugar shack, where the sap gets simmered for hours in shallow pans, boiling away the water until the sugar concentration reaches at least 66 percent. Achieving this requires copious quantities of sap—40 gallons (151 liters) or more to produce a single gallon (3.8 liters) of syrup.

15. The sugar content of honeydew varies widely, depending on the plant species, the type of aphid (or other sap feeder), the

time of year, and even the time of day. The proportions of various sugars in the mix also vary. Fructose, glucose, and sucrose are all common, but so is melezitose, a trisaccharide produced inside the gut of aphids by fusing two units of glucose with one unit of fructose. Melezitose is particularly attractive to ants, and aphids make more of it when ants are present, a pleasing chemical aspect of their fascinating, coevolved relationship. See Fischer and Shingleton 2001.

16. Some have argued for lichens or mushrooms as the biological basis of manna, but neither are common in the Sinai. The ecology of scale insects on tamarisk trees aligns remarkably well with biblical descriptions, from the size of the honeydew pellets to their sweetness to the fact that they often accumulate on the ground overnight. See Bodenheimer 1947 for a fascinating analysis.

17. Furniss and Carolin 1977, p. 96.

18. Hesse 1972, p. 58.

19. Most citizen-driven tree projects rely on smartphones or a good internet connection, but Australian farmer John Nicholas made a remarkable discovery in his back field by employing a different sort of tool altogether: cows. When members of his herd began to suddenly sicken and die, a local veterinarian found their stomachs filled with large, highly toxic seeds unknown to science. They came from a tree standing in Nicholas's pasture that had not been seen anywhere since 1902, and that had never been fully described. Botanists made it the only member of a brand-new genus: *Idiospermum*, a reference to its unique, "idiosyncratic" seeds. Predictably, this resulted in the common name "idiot tree," a useful warning to anyone who might be tempted to sample its poisonous fruits and seeds.

Chapter Six: Below

1. Perhaps nothing illustrates our ancient understanding of plants and soils better than the Chinese character for soil (土), commonly interpreted as topsoil and subsoil (the horizontal bars) with a plant growing up through both. See Gong et al. 2003 for an illuminating history of soil science in China.

2. Hutton 1788, p. 6.

3. From his first paper on the subject, presented to the Geological Society of London in 1837, Darwin enjoyed making the whimsical argument that since "all the vegetable mould over the whole country has passed through, and will again pass through, the intestinal tracts of worms . . . the term 'animal mould' would be in some respects more appropriate" (Darwin 1892, p. 4).

4. The challenge of quantifying species in soil has received a great boost from modern genetic and bioinformatic tools, which can quickly recognize and categorize cryptic organisms like bacteria, fungi, and protists. Even unnamed species can be counted, recognized by unique patterns in their DNA and filed away for later study as "operational taxonomic units." Notably, most estimates do not include phages, the viruses that attack bacteria, a dizzyingly diverse group of taxonomic units that would boost soil diversity numbers even higher. See Anthony et al. 2023, Nielsen et al. 2015, and Roesch et al. 2007 for useful reviews of soil biodiversity.

5. Schlots et al. 1962, p. 44.

6. Peeling back the leaf litter and poking through the top layer of soil usually reveals plenty to look at, particularly if you lie on your belly and peer through a magnifying glass. But Mark Anthony also recommended two sampling methods commonly employed by soil biologists. A Berlese funnel can be constructed from a plastic pop bottle and uses light and heat to drive soil dwellers into a collection jar. (Easy-to-follow instructional videos abound on YouTube.) Also effective, though more disruptive, the mustard extraction relies on the same burning sensation that made hot mustard plasters a popular folk remedy for chest congestion. It involves stirring a generous portion of powdered mustard into a gallon of water, pouring it on the ground, and then gathering up everything that wriggles to the surface as the noxious solution soaks in.

7. The existence of mycorrhizal networks is not in dispute, but there is plenty of scientific controversy about how common they are, how they work, and how they impact the plants and fungi involved. For a good review see Simard et al. 2012, and for a cautionary counterpoint see Karst et al. 2023.

8. When I asked Robert Cichewicz why so many people had jumped at the chance to donate soils to his project, he said that many contributors simply felt a strong commitment to all sorts of citizen science, but a surprising number also mentioned more personal reasons. They had lost loved ones to cancer or other diseases, or were ill themselves, and longed to do something tangible to help researchers find new cures.

9. Schatz 1993, p. 29.

10. Schatz's overlooked role in the discovery of streptomycin led to years of controversy and a major lawsuit over royalties, tainting Waksman's legacy and leaving Schatz embittered. The bacterium at the root of it all, however, fared better. In 2019, by unanimous vote of both legislative chambers, *Streptomyces griseus* became the official microbe of the state of New Jersey.

Chapter Seven: All Wet

1. This passage sometimes appears as "the fairies of my pond," attributable to a misreading of the original French. Monet did not use the word *feeries* (fairies); he said *féeries*, a much richer term that translates to enchantments, magical spectacles, or simply magic.

2. Once considered a variety of the more widespread common snipe (*Gallinago gallinago*), the Wilson's snipe (G. *delicata*) gained full species status based in part on its range (New World versus Old World), but more importantly on differences in its sound-producing tail feathers. Used during elaborate display flights, these plumes vibrate at high speed to make a distinctive winnowing sound involved in marking out territories and attracting mates. For snipes, changing that sound, and the shapes of the feathers that make it, is akin to the evolution of a different song. See Rodrigues et al. 2021.

3. Though not exactly graceful in the air, snipes certainly fly fast, clocking in at 60 miles per hour (96 kilometers per hour) and known for making sudden sharp turns and rapid, midair zigzags. Hunters require highly accurate aim to bring one down on the wing, an association at the root of the military term "sniper."

4. Teale 1974, p. 137.

5. Ibid.

6. Ibid.

7. Ibid.

8. This process eventually resulted in several gorilla families being willing to accept small groups of daily visitors, the basis for an ongoing ecotourism program that continues to help sustain Uganda's Bwindi Impenetrable National Park. See Hanson 2014 for a detailed exploration of a very different sort of backyard ecosystem!

9. Just as the sounds filtering into a stick pile inspire new reactions, so too do the sounds emanating outward. In one of my favorite of Teale's anecdotes, he described listening to a flock of forty grackles foraging on and around his hideaway while trying in vain to hold back a sneeze. When it finally came, the sudden, disembodied "Achoo!" shocked the birds into panicky flight with "a harsh clamor and an airy roaring of wings" (Teale 1974, p. 141).

10. Technically, the distinction has to do with metamorphosis. Roughly three-quarters of insects go through "complete metamorphosis," with four distinct life stages: egg, larva, pupa, and adult. In these species, the larvae are specialized for feeding and do not resemble the adults (e.g., maggot versus fly; grub versus beetle). The remaining insects, including dragonflies and other familiar groups like grasshoppers and cicadas, have larval stages more similar to the adults in appearance and habit. They mature through "gradual metamorphosis," a process consisting of only three stages: egg, nymph, and adult.

11. van Leeuwenhoek 1702, p. 1304.

12. Ibid., p. 1305.

13. Ibid.

14. Ibid., p. 1306.

15. Ibid.

16. van Leeuwenhoek 1979, p. 155.

17. Ibid., p. 129.

18. Ibid., p. 127.

19. The key to analyzing eDNA lies in a process called metabarcoding that focuses on particular, highly variable lengths of genetic

code shared by all species. Small differences in that code are unique to different organisms, so if they are present in a sample of water or soil, or even in the air, it follows that the organism must have been present too. Microbiologists have been doing something similar for years to inventory bacteria, which would otherwise need to be painstakingly grown in cultures on petri dishes to be identified. *Macro*-biologists have caught on relatively recently, with remarkable results. See Ruppert et al. 2019 for a review.

Chapter Eight: After Hours

1. General estimates of owl strike force come from studies of a captive barn owl trained to strike at a pressure plate. Barred owls hunt similar prey with similar tactics, relying on a powerful strike and sustained gripping strength to immobilize the small mammals they favor. The power of their strike also helps them plunge through thick grasses or deep snow cover. See Usherwood et al. 2014.

2. Few true crime stories have received as much media attention as the murder of Kathleen Peterson and the controversial conviction of her husband, who later walked free. The barred owl theory features prominently in many of the dozen or so films, books, documentaries, podcasts, and TV episodes dedicated to the case.

3. Bent 1938, p. 184.

4. The observations reported by Bent probably included some unrecorded turnover in those breeding pairs—barred owls are thought to rarely live longer than ten to fifteen years in the wild. Still, a thirty-four-year span on a territory is theoretically possible. At least three individuals living in captivity have approached or exceeded that venerable age. See Orfinger et al. 2018.

5. Perhaps the real knowledge deficit lies in our appreciation of "crepuscular biology": only 652,000 mentions on the entire World Wide Web!

6. Extinction rates for large African carnivores rose in apparent lockstep with increased brain size for hominins, and studies suggest that the two trends may be related. Higher intelligence, tool use, and complex social coordination would have helped our ancestors steal kills from other predators, as well as compete with them for

limited ungulate prey. There is also evidence that hominins hunted large carnivores directly—for their meat as well as for their significance in cultural rituals. See Faurby et al. 2020 and Russo et al. 2023 for compelling theories about what may be one of the earliest discernable human impacts on biodiversity.

7. Controlled human use of fire was widespread by 400,000 years ago, and some experts believe the practice became established much earlier. Evidence from a site in modern-day Israel associated with *Homo erectus*, for example, suggests that members of our genus have been building campfires (and presumably huddling around them) for at least 790,000 years. See MacDonald et al. (2021) for a thoughtful review.

8. Atavistic threat responses can be innate and immediate, but they can also manifest in how easily fears are acquired. In a famous study of rhesus monkeys, for example, naïve, laboratory-reared individuals quickly learned to fear rubber snakes by watching the reactions of more experienced, wild-reared monkeys. No amount of similar conditioning, however, could teach them to fear a flower, or a stuffed bunny. See Cook and Mineka 1989, 1990.

9. Once discovered, the wren nest's convenient location made for easy daylight observations of the birds that called it home. I was at La Selva for other reasons, but by pausing en route between the dormitory and various research obligations, I managed to spend 403 delightful minutes watching stripe-breasted wrens incubate their eggs. See Hanson 2006.

10. Though American inventor Thomas Edison is often credited with the invention of the lightbulb, many others were working on similar designs at the time. In England, Joseph Swan patented his version within months of Edison's. The two soon joined forces, bringing Swan's ideas to what would become General Electric in America, and also combining in Britain to form the more poetically named company Ediswan.

11. While many species remain stymied by artificial light, some have responded with distinct behavioral changes, ranging from outright avoidance to altered sleep patterns to new predation strategies (i.e., eating things disoriented by lightbulbs). In a few cases, there

are even signs of measurable evolution taking place. Urban populations of the small ermine moth (*Yponomeuta cagnagella*), for example, are now less attracted to light sources than their rural, dark-sky counterparts. Because flying toward artificial lights results in higher mortality, natural selection has presumably begun winnowing that impulse from the "city moths," either directly, or by selecting for less visual perception and/or mobility in general. See Altermatt and Ebert 2016 and also Hopkins et al. 2018, who argue that artificial light is an underappreciated force in the evolution of distinct urban populations for all manner of organisms.

12. Satellite measurements suggest the extent and intensity of outdoor lighting is growing by roughly 2 percent annually, but a recent ground-based study from citizen science observations put the rate of annual brightening at close to 10 percent. In either case, artificial sky glow has already made many constellations and the glowing band of the Milky Way invisible to an estimated 80 percent of people in urbanized countries. Stargazers lament the loss of such a basic touchstone of human inspiration, but it impacts other species too. Various creatures from birds to sea turtles use the stars to navigate, and African dung beetles use the Milky Way itself, orienting their rolling activities in relation to its distinct, linear pattern. The loss of that nighttime beacon and other stellar landmarks probably contributes to declines of beetles and many other groups in developed landscapes. See Kyba et al. 2017, 2023 and Dacke et al. 2013.

13. Schroer and colleagues from across Europe and North America recently called for the establishment of "nyctology," from the Greek for "night study," an interdisciplinary field of science dedicated to understanding what happens after dark, and how those essential processes are threatened by artificial light. See Kyba et al. 2020 and also Gaston 2019.

14. Sibylle Schroer recommends avoiding altogether any bulbs that emit a cold, white light. Those above 3000 on the Kelvin color temperature scale shine with too much light in the blue portion of the spectrum, the most harmful glow for the majority of species (including *Homo sapiens*).

15. Kevin Gaston, an artificial light expert at the University of Exeter, reminded me of an even simpler technique for reducing backyard light pollution: drawing the blinds.

16. Fighting light pollution can take other forms too, from cell phone apps like Globe at Night, which helps scientists measure sky glow around the world, to organizations like DarkSky International, which promotes smart lighting, education, and neighbor-to-neighbor activism in more than thirty countries.

Chapter Nine: The Welcome Mat

1. Ever since that incident, I have racked my brain to think of another phrase that has the opposite meaning when read upside down. If you come up with one, let me know!

2. Biographers note that Alfred Kinsey lacked confidence with sampling design and statistical analysis, a scientific shortcoming he overcame in all of his research by amassing prodigious amounts of data. For his studies on human sexuality, he personally participated in 7,985 interviews, and his colleagues contributed roughly 10,000 more. See Jones 2004 for an exhaustive account of his life and work.

3. Curators at the American Museum of Natural History describe how Kinsey's abrupt transition to sex research left many gall wasp projects incomplete—new species half-described, taxonomies half-revised, manuscripts unfinished, etc. But the sheer scope of the collection, and its potential for additional study, make it "one of the greatest assets" of the entire museum.

4. Kinsey's research focused on the diverse galls associated with oak trees, but he also collected widely from other plants. On a trip through the American West in 1920 he gathered thimbleberry galls from California north to the Canadian border. He used that material to suggest splitting the gall maker into several subspecies, a taxonomic revision that remains unresolved. See Kinsey 1922.

5. Mimicry in plant galls borders on the mind-bending. Some aphids in the genus *Tuberocephalus*, for example, produce galls on the edges of cherry leaves that look just like plump, pinkish caterpillars,

complete with fake body segments. But that's not all—the galls also curl in a way that makes the leaves look scalloped and chewed, as if the "caterpillar" had been feeding. This deception reduces real caterpillar nibbling because many moths and butterflies avoid laying eggs on leaves where a competitor is already established. (They want their offspring to have a leaf all to themselves.) For the gall maker, it means a healthier, non-chewed leaf with more energy to invest in making gall tissue. And for evolutionary biologists, it means a bizarre, multitiered system where the genes of the gall maker are responding to the behavior of moths and butterflies to reshape and fine-tune the appearance of the plant. For all intents and purposes, the leaves of the cherry trees become like an extension of the aphid's body! See Yamazaki 2016.

6. Kinsey 1920, p. 319.

7. Most fleabanes belong to the genus *Erigeron*, a large group in the aster family with over 450 varieties scattered primarily across Europe, Asia, and the Americas. The name refers to the apparent role of these plants in traditional medicine as a flea repellent, or even a flea killer. But while extracts from several *Erigeron* species have shown promise for their antifungal and anti-inflammatory properties, their effectiveness as a bane to fleas has yet to be scientifically confirmed.

8. At least two gall-forming midges also colonized that first fleabane in Blyth and Eiseman's lawn, and various bees and flies were spotted visiting its flowers. As Eiseman later noted in a blog post, the plant was definitely "earning its keep"!

9. In a follow-up exchange of emails, Eiseman gave me an update on the yard's leaf miner total: "Let's see . . . 15 beetles, 89 flies, 10 sawflies, 120 moths. . . . At least 234!" Among those, three of the flies were new to science. In taxonomic circles, that means that Eiseman and Blyth's yard will forever be known as the "type locality" for those species, the place where the first specimens were collected.

10. Sawflies and other leaf eaters include many species regarded as crop and garden pests, so welcoming them into the yard can

require a bit of an attitude adjustment. Most are completely harmless, even if their activity looks dramatic. One year, for example, Eiseman and Blyth's spice bushes were completely denuded by the caterpillars of a swallowtail butterfly. But the bushes fully recovered, leafing out again the next spring and continuing to thrive. "Things will eat plants, and that's OK," Eiseman concluded. "They'll be fine. They've evolved for that."

11. Though trees festooned with butterflies are spectacular to witness, the habit of concentrating so many individuals in such small places makes monarch populations vulnerable to everything from wildfires and windstorms to climate change and deforestation. To put that risk into perspective, the entire eastern North American subspecies overwinters in Mexican forests that total less than 5 acres (2 hectares). That's about the size of an average shopping mall parking lot.

12. Like Emily Geest, most experts caution against extrapolating the results of any one backyard biodiversity study. The quality of urban and suburban habitats varies widely, as does their context within broader, landscape-level systems. But there is mounting evidence that small habitat patches can have cumulative effects at larger scales. For thoughtful examples, see Grade et al. 2022, Donkersley et al. 2023, and related references therein.

Chapter Ten: The Limiting Factor

1. Remarkably, ornithologist Waldo Lee McAtee considered this list to be woefully incomplete! See McAtee 1926, p. 84.

2. Historians note that Liebig's fame as a founder of organic chemistry allowed him to appropriate Sprengel's idea with little consequence, even though the two were contemporaries and Sprengel protested the injustice. Recent attempts to rename the idea as "Sprengel-Liebig's Law" have met with little success. See van der Ploeg et al. 1999 for a thoughtful review of both men's contributions.

3. Brewer 1879, p. 88.

4. Ibid.

5. Ibid.

6. Ibid., p. 89.

7. Should you ever find yourself trying to identify a brown creeper nest, look for strips of shredded bark amidst an extremely eclectic jumble of materials. Various wrens, which might nest in similar locations, are generally twiggier in their habits and do not use shredded bark.

8. Tyler 1914, p. 51.

9. Metaphorically, Joe Buckler's professional background and biological contributions come together perfectly in the Spanish term for woodpecker: *el carpintero*, "the carpenter," a reference not only to the birds' home-building skills, but to the way they fill the woods with the sounds of tapping and hammering.

10. While attaching precise census data to population trends is difficult, Benedikt Schmidt is confident that overall numbers are rising for most pond-breeding amphibians in Aargau. As a proxy, he and his colleagues analyzed data for male European tree frogs (*Hyla arborea*), the easiest group in the study to detect and count. "The numbers went up substantially," he told me. "More individuals, more populations. There are definitely more frogs now than in the 1990s." In the lexicon of conservation and ecological restoration, that earns male tree frogs in Aargau a rarely used label: "beyond recovery."

11. The reason for this pattern is rooted in the availability of certain omega-3 fatty acids, known to many health-conscious eaters as "the good fats." Because omega-3s are produced more easily by plants in aquatic environments, they accumulate in aquatic food webs, and insects that go through their larval phase in ponds have them in abundance. Most terrestrial plants lack omega-3s, and so do the insects that feed on them. For swallow chicks, eating a diet rich in omega-3s has been shown to improve growth rates, body condition, and immunocompetence. See Twining et al. 2018 for a fascinating exploration of this topic.

12. There is a small but growing body of scientific literature dedicated to testing the broader impacts of small-scale habitat projects. The examples listed here include encouraging cumulative effects of pollinator gardens in Washington State (Donkersley et al. 2023),

the diversification and growth of feeder-visiting bird populations in Britain (Plummer et al. 2019), and a rapid uptick in hoopoes in central Europe following a mass distribution of nest boxes (Berthier et al. 2012).

Conclusion: Wild Crescendo

1. Beumer and Martens 2015.

2. Cells from the caterpillars of the fall armyworm moth (*Spodoptera frugiperda*) have become common protein-multipliers in vaccine production. The soapbark tree (*Quillaja saponaria*) provides the adjuvant, a substance that increases the body's immune response and makes the vaccine more effective. See Wadman 2020 for a fascinating history of the process.

3. In an era of climate change, the value of well-vegetated yards is likely to rise right alongside global temperatures. Studies in California and Alabama suggest that the cooling moisture provided by backyard tree cover reduces air-conditioning expenses by anywhere from 9 to 30 percent. See Akbari et al. 1997 and Pandit and Laband 2010 for discussions of cooling, and Anderson and Cordell 1988, Des Rosiers et al. 2002, and Dimke et al. 2013 for real estate examples.

4. Biologists call this sort of effect a "trophic cascade," where changes ripple through the many interconnected levels of a food web, from predators to their herbivore prey, to the plants they feed upon, and so on.

5. Among many examples of controversy, the importance of Yellowstone wolf reintroduction to the recovery of aspen trees has been both challenged (Brice et al. 2022) and vigorously defended (Beschta et al. 2023), as has the role of megafaunal extinctions versus climate perturbations in changing Australia's vegetation (Rule et al. 2012, Cooper et al. 2021). When one doesn't have an academic axe to grind, it's clear that such debates boil down to matters of degree—there is widespread agreement that extinctions and reintroductions impact ecosystems; the question is by how much.

6. See Schmitz et al. 2023 for a fascinating discussion about how "trophic diversity" increases carbon sequestration.

7. Ulrich 1984.

8. See Frumkin et al. 2017, Aerts et al. 2018, and Nguyen et al. 2023 for excellent reviews of the many health benefits of nature exposure.

9. Aerts et al. 2018.

Bibliography

Abramson, C. I. 2006. Charles Henry Turner: Pioneer of comparative psychology. In *Portraits of Pioneers in Psychology, Volume VI*, edited by D. A. Dewsbury, L. T. Benjamin Jr., and M. Wertheimer, 37–49. Sussex: Psychology Press.

Abramson, C. I. 2009. A study in inspiration: Charles Henry Turner (1867–1923) and the investigation of insect behavior. *Annual Review of Entomology* 54: 343–359.

Aerts, R., O. Honnay, and A. Van Nieuwenhuyse. 2018. Biodiversity and human health: Mechanisms and evidence of the positive health effects of diversity in nature and green spaces. *British Medical Bulletin* 127(1): 5–22.

Agersnap, S., E. E. Sigsgaard, M. R. Jensen, M. D. P. Avila, H. Carl, P. R. Møller, S. L. Krøs, S. W. Knudsen, M. S. Wisz, and P. F. Thomsen. 2022. A national scale "BioBlitz" using citizen science and eDNA metabarcoding for monitoring coastal marine fish. *Frontiers in Marine Science* 9: 824100.

Akbari, H., D. M. Kurn, S. E. Bretz, and J. W. Hanford. 1997. Peak power and cooling energy savings of shade trees. *Energy and Buildings* 25(2): 139–148.

Altermatt, F., and D. Ebert. 2016. Reduced flight-to-light behaviour of moth populations exposed to long-term urban light pollution. *Biology Letters* 12(4): 20160111.

Anderson, L. M., and H. K. Cordell. 1988. Influence of trees on residential property values in Athens, Georgia (U.S.A.): A survey based on actual sales prices. *Landscape and Urban Planning* 15(1–2): 153–164.

Anonymous (Samuel Scudder). 1874. In the laboratory with Agassiz. *Every Saturday: A Journal of Choice Reading (New Series)* 1: 369–370.

Bibliography

Anthony, M. A., S. F. Bender, and M. G. van der Heijden. 2023. Enumerating soil biodiversity. *Proceedings of the National Academy of Sciences* 120(33): e2304663120.

Aplin, L. M., R. E. Major, A. Davis, and J. M. Martin. 2021. A citizen science approach reveals long-term social network structure in an urban parrot, *Cacatua galerita*. *Journal of Animal Ecology* 90(1): 222–232.

Archer, W. 1915. *Fighting a Philosophy*. Oxford Pamphlets 1914–1915. London: Oxford University Press.

Aston, W. G. 1905. *A History of Japanese Literature*. New York: D. Appleton and Company.

Atkins, J. L., R. A. Long, J. Pansu, J. H. Daskin, A. B. Potter, M. E. Stalmans, C. E. Tarnita, and R. M. Pringle. 2019. Cascading impacts of large-carnivore extirpation in an African ecosystem. *Science* 364(6436): 173–177.

Balosso-Bardin, C., A. Ernoult, P. de la Cuadra, B. Fabre, and I. Franciosi. 2018. The Secret of the bagpipes: Controlling the bag. Techniques, skill and musicality. *Galpin Society Journal* 7: 189–272.

Baril, L. M., A. J. Hansen, R. Renkin, and R. Lawrence. 2011. Songbird response to increased willow (*Salix* spp.) growth in Yellowstone's northern range. *Ecological Applications* 21(6): 2283–2296.

Bates, A. E., R. B. Primack, B. S. Biggar, T. J. Bird, M. E. Clinton, R. J. Command, C. Richards, et al. 2021. Global COVID-19 lockdown highlights humans as both threats and custodians of the environment. *Biological Conservation* 263: 109175.

Beck, B. B. 1980. *Animal Tool Use: The Use and Manufacture of Tools by Animals*. New York: Garland.

Beggs, J. 2001. The ecological consequences of social wasps (*Vespula* spp.) invading an ecosystem that has an abundant carbohydrate resource. *Biological Conservation* 99(1): 17–28.

Beggs, J. R., and D. A. Wardle. 2006. Keystone species: Competition for honeydew among exotic and indigenous species. In *Biological Invasions in New Zealand*, edited by R. B. Allen and W. G. Lee, 281–294. Berlin: Springer.

Bent, A. C. 1938. *Life Histories of North American Birds of Prey, Part Two*. New York: Dover.
Bent, A. C. 1948. *Life Histories of North American Nuthatches, Wrens, Thrashers, and Their Allies*. New York: Dover.
Berthier, K., F. Leippert, L. Fumagalli, and R. Arlettaz. 2012. Massive nest-box supplementation boosts fecundity, survival and even immigration without altering mating and reproductive behaviour in a rapidly recovered bird population. *PLoS ONE* 7(4): e36028.
Beschta, R. L., L. E. Painter, and W. J. Ripple. 2023. Revisiting trophic cascades and aspen recovery in northern Yellowstone. *Food Webs* 36: e00276.
Beschta, R. L., and W. J. Ripple. 2012. Berry-producing shrub characteristics following wolf reintroduction in Yellowstone National Park. *Forest Ecology and Management* 276: 132–138.
Beschta, R. L., and W. J. Ripple. 2018. Can large carnivores change streams via a trophic cascade? *Ecohydrology* 12(1): e2048.
Beumer, C., and P. Martens. 2015. Biodiversity in my (back) yard: Towards a framework for citizen engagement in exploring biodiversity and ecosystem services in residential gardens. *Sustainability Science* 10(1): 87–100.
Bevington, D., ed. 1980. *The Complete Works of Shakespeare*. Glenview, IL: Scott, Foresman and Company.
Bodenheimer, F. S. 1947. The manna of Sinai. *Biblical Archaeologist* 10(1): 1–6.
Bolliger, J., T. Hennet, B. Wermelinger, R. Bösch, R. Pazur, S. Blum, J. Haller, et al. 2020. Effects of traffic-regulated street lighting on nocturnal insect abundance and bat activity. *Basic and Applied Ecology* 47: 44–56.
Bostock, J., and H. T. Riley, trans. 1855. *The Natural History of Pliny, Volume III*. London: Henry G. Bohn.
Brevik, E. C., and A. E. Hartemink. 2010. Early soil knowledge and the birth and development of soil science. *Catena* 83(1): 23–33.
Brewer, R. 1961. Comparative notes on the life history of the Carolina Chickadee. *The Wilson Bulletin* 73(4): 348–373.
Brewer, T. M. 1879. The American brown creeper. *Bulletin of the Nuttall Ornithological Club* 4(2): 87–90.

Brewster, W. 1879. Breeding habits of the American brown creeper (*Certhia familiaris americana*). *Bulletin of the Nuttall Ornithological Club* 4(4): 199–209.

Brice, E. M., E. J. Larsen, and D. R. MacNulty. 2022. Sampling bias exaggerates a textbook example of a trophic cascade. *Ecology Letters* 25(1): 177–188.

Brodo, I. M., S. D. Sharnoff, and S. Sharnoff. 2001. *Lichens of North America*. New Haven, CT: Yale University Press.

Brown, B. V. 2012. Small size no protection for acrobat ants: World's smallest fly is a parasitic phorid (Diptera: Phoridae). *Annals of the Entomological Society of America* 105(4): 550–554.

Brown, C. R., and M. B. Brown. 2013. Where has all the road kill gone? *Current Biology* 23(6): R233–R234.

Burkhardt, R. W., Jr. 2005. *Patterns of Behavior: Konrad Lorenz, Nikon Tinbergen, and the Founding of Ethology*. Chicago: University of Chicago Press.

Burroughs, J. 1908. *Leaf and Tendril*. Boston: Houghton Mifflin.

Burroughs, J. 1910. *In the Catskills*. Boston: Houghton Mifflin.

Buxton, A., A. Diana, E. Matechou, J. Griffin, and R. A. Griffiths. 2022. Reliability of environmental DNA surveys to detect pond occupancy by newts at a national scale. *Scientific Reports* 12(1): 1295.

Ceríaco, L. M. P., B. S. Santos, M. P. Marques, A. M. Bauer, and A. Tiutenko. 2021. Citizen science meets specimens in old formalin filled jars: A new species of banded rubber frog, genus *Phrynomantis* (Anura: Phrynomeridae) from Angola. *Alytes* 38(1–4): 18–48.

Cervantes Saavedra, M. 1916. *The Ingenious Gentleman Don Quixote of La Mancha, Vols. III and IV*. Translated by J. Ormsby. New York: Dodd, Mead and Company.

Chen, L., and S. D. Porter. 2020. Biology of *Pseudacteon* decapitating flies (Diptera: Phoridae) that parasitize ants of the *Solenopsis saevissima* complex (Hymenoptera: Formicidae) in South America. *Insects* 11(2): 107.

Cheng, W., R. C. Kendrick, F. Guo, S. Xing, M. W. Tingley, and T. C. Bonebreak. 2019. Complex elevational shifts in a tropical lowland moth community following a decade of climate change. *Diversity and Distributions: Special Issue: Conservation Biogeography in a Changing Climate* 25(4): 514–523.

Cook, M., and S. Mineka. 1989. Observational conditioning of fear to fear-relevant versus fear-irrelevant stimuli in rhesus monkeys. *Journal of Abnormal Psychology* 98(4): 448–459.

Cook, M., and S. Mineka. 1990. Selective associations in the observational conditioning of fear in rhesus monkeys. *Journal of Experimental Psychology: Animal Behavior Processes* 16(4): 372–389.

Cooper, A., C. S. M. Turney, J. Palmer, A. Hogg, M. McGlone, J. Wilmshurst, A. Lorrey, et al. 2021. A global environmental crisis 42,000 years ago. *Science* 371(6531): 811–818.

Costa, J. T. 2017. *Darwin's Backyard.* New York: W. W. Norton and Company.

Dacke, M., E. Baird, M. Byrne, C. H. Scholtz, and E. J. Warrant. 2013. Dung beetles use the Milky Way for orientation. *Current Biology* 23(4): 298–300.

Darwin, C. 1892. *The Formation of Vegetable Mould, Through the Action of Worms, with Observations on Their Habits.* London: John Murray.

Darwin, C. 2004. *The Voyage of the Beagle* (Reprint of 1845 edition). Washington, DC: National Geographic Society.

Darwin Correspondence Project. 1855. "Letter no. 1693." Accessed December 9, 2022. www.darwinproject.ac.uk/letter/?docId=letters/DCP-LETT-1693.xml.

Darwin, C. R., ed. 1838. *Birds Part 3 No. 1 of The zoology of the voyage of H.M.S. Beagle. by John Gould. Edited and superintended by Charles Darwin.* London: Smith Elder and Co.

Darwin, F., ed. 1908. *Charles Darwin: His Life Told in an Autobiographical Chapter and in a Selected Series of His Published Letters.* London: John Murray.

Darwin, F., L. Darwin, and H. Darwin. 1859. Coleoptera at Down. *Entomologist's Weekly Intelligencer* 6: 99.

Debrot, A. O. 2014. Nocturnal foraging by artificial light in three Caribbean bird species. *Journal of Caribbean Ornithology* 27: 40–41.

Denison, J. C. 1973. Life in tall trees. *Scientific American* 228(6): 74–80.

de Pennart, A., and P. G. Matthews. 2020. The bimodal gas exchange strategies of dragonfly nymphs across development. *Journal of Insect Physiology* 120: 103982.

Derryberry, E. P., J. N. Phillips, G. E. Derryberry, M. J. Blum, and D. Luther. 2020. Singing in a silent spring: Birds respond to a half-century soundscape reversion during the COVID-19 shutdown. *Science* 370(6516): 575–579.

Des Rosiers, F., M. Thériault, Y. Kestens, and P. Villeneuve. 2002. Landscaping and house values: An empirical investigation. *Journal of Real Estate Research* 23(1/2): 139–162.

Dillard, A. 1989. *A Writing Life*. New York: Harper and Row.

Dimke, K. C., T. D. Sydnor, and D. S. Gardner. 2013. The effect of landscape trees on residential property values of six communities in Cincinnati, Ohio. *Arboriculture and Urban Forestry* 39(2): 49–55.

Donkersley, P., S. Witchalls, E. H. Bloom, and D. W. Crowder. 2023. A little does a lot: Can small-scale planting for pollinators make a difference? *Agriculture, Ecosystems & Environment* 343: 108254.

Doolittle, E. L. 2020. "Hearken to the Hermit-Thrush": A case study in interdisciplinary listening. *Frontiers in Psychology* 11: 613510.

Doolittle, E. L., B. Gingras, D. M. Endres, and W. T. Fitch. 2014. Overtone-based pitch selection in hermit thrush song: unexpected convergence with scale construction in human music. *Proceedings of the National Academy of Sciences* 111: 16616–16621.

Driessen, M. 2021. COVID-19 restrictions provide a brief respite from the wildlife roadkill toll. *Biological Conservation* 256: 109012.

Du, L., A. J. Robles, J. B. King, D. R. Powell, A. N. Miller, S. L. Mooberry, and R. H. Cichewicz. 2014. Crowdsourcing natural products discovery to access uncharted dimensions of fungal metabolite diversity. *Angewandte Chemie International Edition* 53(3): 804–809.

Du Bois, W. E. B. 1939. The negro scientist. *The American Scholar* 8(3): 309–320.

Durand, G. A., D. Raoult, and G. Dubourg. 2019. Antibiotic discovery: History, methods and perspectives. *International Journal of Antimicrobial Agents* 53(4): 371–382.

Egan, S. P., G. R. Hood, E. O. Martinson, and J. R. Ott. 2018. Cynipid gall wasps. *Current Biology* 28(24): R1370–R1374.

Elder, M. 1924. *À Giverny, chez Claude Monet*. Paris: Bernheim-Jeune.

Eliot, G. 1880. *Middlemarch: A Study of Provincial Life*. London: William Blackwood and Sons.

Erwin, T. L. 1982. Tropical forests: Their richness in Coleoptera and other arthropod species. *The Coleopterists Bulletin* 36(1): 74–75.

Erwin, T. L. 1983. Tropical forest canopies: The last biotic frontier. *Bulletin of the Ecological Society of America* 29: 14–20.

Erwin, T. L. 1988. The tropical forest canopy: The heart of biotic diversity. In *Biodiversity*, edited by E. O. Wilson, 123–129. Washington, DC: National Academy Press.

Erwin, T. L., and J. C. Scott. 1980. Seasonal size patterns, trophic structure, and richness of Coleoptera in the tropical arboreal ecosystem: The fauna of the tree *Luhea seemannii* Triana and Planch in the Canal Zone of Panama. *The Coleopterists Bulletin* 34(2): 305–322.

Ethier, J. P., A. Fayard, P. Soroye, D. Choi, M. J. Mazerolle, and V. L. Trudeau. 2021. Life history traits and reproductive ecology of North American chorus frogs of the genus *Pseudacris* (Hylidae). *Frontiers in Zoology* 18: 40.

Fabre, J. H. 1914. *The Mason Bees*. New York: Dodd, Mead and Company.

Fabre, J. H. 1919. *The Mason-wasps*. New York: Dodd, Mead and Company.

Fagan, M. M. 1918. The uses of insect galls. *The American Naturalist* 52: 155–176.

Fagan-Jeffries, E. P., and A. D. Austin. 2021. Four new species of parasitoid wasp (Hymenoptera: Braconidae) described through a citizen science partnership with schools in regional South Australia. *Zootaxa* 4949(1): 79–101.

Farvet, C. 1999. Jean-Henri Fabre: His life experiences and predisposition against Darwinism. *American Entomologist* 45(1): 38–48.

Faurby, S., D. Silvestro, L. Werdelin, and A. Antonelli. 2020. Brain expansion in early hominins predicts carnivore extinctions in East Africa. *Ecology Letters* 23(3): 537–544.

Fern, F. 1870. *Ginger-snaps*. New York: Carlton.

Fischer, M. K., and A. W. Shingleton. 2001. Host plant and ants influence the honeydew sugar composition of aphids. *Functional Ecology* 15(4): 544–550.

Fontaine, B., K. van Achterberg, M. A. Alonso-Zarazaga, R. Araujo, M. Asche, H. Aspöck, U. Aspöck, et al. 2012. New species in the Old World: Europe as a frontier in biodiversity exploration, a test bed for 21st century taxonomy. *PLoS ONE* 7(5): e36881.

Fontaine, C., B. Fontaine, and A. C. Prévot. 2021. Do amateurs and citizen science fill the gaps left by scientists? *Current Opinion in Insect Science* 46: 83–87.

Forister, M. L., C. A. Halsch, C. C. Nice, J. A. Fordyce, T. E. Dilts, J. C. Oliver, K. L. Prudic, et al. 2021. Fewer butterflies seen by community scientists across the warming and drying landscapes of the American West. *Science* 371(6533): 1042–1045.

Fournet, M. 2021. The impact of the "Anthropause" on the communication and acoustic habitat of Southeast Alaskan humpback whales. Acoustical Society of America Acoustics in Focus Virtual Meeting, June 8–10.

Fowler, D. W., E. A. Freedman, and J. B. Scannella. 2009. Predatory functional morphology in raptors: Interdigital variation in talon size is related to prey restraint and immobilisation technique. *PLoS ONE* 4(11): e7999.

Friedlingstein, P., M. O'Sullivan, M. W. Jones, R. M. Andrew, J. Hauck, A. Olsen, G. P. Peters, et al. Global carbon budget 2020. *Earth System Science Data* 12(4): 3269–3340.

Frumkin, H., G. N. Bratman, S. J. Breslow, B. Cochran, P. H. Kahn Jr., J. J. Lawler, P. S. Levin, et al. 2017. Nature contact and human health: A research agenda. *Environmental Health Perspectives* 125(7): 075001.

Furniss, R. L., and V. M. Carolin. 1977. *Western Forest Insects. Miscellaneous Publication No. 1339*. Washington, DC: US Department of Agriculture, Forest Service.

Gaston, K. J. 2019. Nighttime ecology: The "nocturnal problem" revisited. *The American Naturalist* 193(4): 481–502.

GBIF (Global Biodiversity Information Facility). 2022. What is GBIF? Archived at www.gbif.org/what-is-gbif. Accessed October 27, 2022.

Geest, E. A., L. L. Wolfenbarger, and J. P. McCarty. 2019. Recruitment, survival, and parasitism of monarch butterflies (*Danaus plexippus*) in milkweed gardens and conservation areas. *Journal of Insect Conservation* 23: 211–224.

Gest, H. 2004. The discovery of microorganisms by Robert Hooke and Antoni van Leeuwenhoek, fellows of the Royal Society. *Notes and Records of the Royal Society of London* 58(2): 187–201.

Goldstein, E. B., E. V. Mullins, L. J. Moore, R. G. Biel, J. K. Brown, S. D. Hacker, K. R. Jay, et al. 2018. Literature-based latitudinal distribution and possible range shifts of two US east coast dune grass species (*Uniola paniculata* and *Ammophila breviligulata*). *PeerJ* 6: e4932.

Gong, Z., X. Zhang, J. Chen, and G. Zhang. 2003. Origin and development of soil science in ancient China. *Geoderma* 115: 3–13.

Goodenough, A. E., N. Little, W. S. Carpenter, and A. G. Hart. 2017. Birds of a feather flock together: Insights into starling murmuration behaviour revealed using citizen science. *PLoS ONE* 12(6): e0179277.

Grade, A. M., P. S. Warren, and S. B. Lerman. 2022. Managing yards for mammals: Mammal species richness peaks in the suburbs. *Landscape and Urban Planning* 220: 104337.

Grahame, K. 1920. *The Wind in the Willows*. New York: Charles Scribner's Sons.

Grieneisen, M. L., Y. U. Zhan, D. Potter, and M. Zhang. 2014. Biodiversity, taxonomic infrastructure, international collaboration, and new species discovery. *BioScience* 64(4): 322–332.

Hanson, T. 2006. First observations of incubation behavior for the stripe-breasted wren (*Thryothorus thoracius*). *Ornitología Neotropical* 17(3): 453–456.

Hanson, T. 2014. *The Impenetrable Forest: Gorilla Years in Uganda*. New York: Curtis Brown.

Hanson, T. 2022. Predation of a Pacific tree frog (*Pseudacris regilla*) by an American robin (*Turdus migratorius*) on San Juan Island, Washington. *Northwestern Naturalist* 103(2): 190–193.

Harris, S. E., and J. Munshi-South. 2017. Signatures of positive selection and local adaptation to urbanization in white-footed mice (*Peromyscus leucopus*). *Molecular Ecology* 26(22): 6336–6350.

Harrison, S. E., S. J. Oliver, D. S. Kashi, A. T. Carswell, J. P. Edwards, L. M. Wentz, R. Roberts, et al. 2021. Influence of vitamin D supplementation by simulated sunlight or oral D3 on respiratory infection during military training. *Medicine and Science in Sports and Exercise* 53(7): 1505–1516.

Hart, A. G., W. S. Carpenter, E. Hlustik-Smith, M. Reed, and A. E. Goodenough. 2018. Testing the potential of Twitter mining methods for data acquisition: Evaluating novel opportunities for ecological research in multiple taxa. *Methods in Ecology and Evolution* 9(11): 2194–2205.

Harvey, A. G. 1947. *Douglas of the Fir*. Cambridge, MA: Harvard University Press.

Herr, H., S. Viquerat, F. Devas, A. Lees, L. Wells, B. Gregory, T. Giffords, et al. 2022. Return of large fin whale feeding aggregations to historical whaling grounds in the Southern Ocean. *Scientific Reports* 12: 9458.

Hesse, H. 1972. *Wandering*. Translated by J. Wright. New York: Farrar, Straus and Giroux.

Holland, P., trans. 1601. *The Historie of the World: Commonly Called the Natural Historie of C. Plinius Secundus*. London: Adam Islip.

Hopkins, G. R., K. J. Gaston, M. E. Visser, M. A. Elgar, and T. M. Jones. 2018. Artificial light at night as a driver of evolution across urban–rural landscapes. *Frontiers in Ecology and the Environment* 16(8): 472–479.

Horace. 1926. *Satires, Epistles, and Ars Poetica*. Translated by W. R. Faircloth. New York: G. P. Putnam's Sons.

Hutton, J. 1788. Theory of the earth. *Transactions of the Royal Society of Edinburgh* 1: 209–304.

Hyde, K. D., J. Xu, S. Rapior, R. Jeewon, S. Lumyong, A. G. T. Niego, P. D. Abeywickrama, et al. 2019. The amazing potential of fungi: 50 ways we can exploit fungi industrially. *Fungal Diversity* 97: 1–136.

Irwin, A. 1995. *Citizen Science*. London: Routledge.

Jain, P., H. Forbes, and L. A. Esposito. 2022. Two new alkali-sink specialist species of *Paruroctonus* Werner 1934 (Scorpiones, Vaejovidae) from central California. *ZooKeys* 1117: 139–188.

James, D. G., and D. Nunnallee. 2011. *Life Histories of Cascadia Butterflies*. Corvallis, OR: Oregon State University Press.

Johnson, S. 1787. *The Works of Samuel Johnson, LL.D., vol. 11*. London: J. Buckland.

Jones, J. H. 2004. *Alfred C. Kinsey: A Life*. New York: W. W. Norton and Company.

Kaplan, S. 1995. The restorative benefits of nature: Toward an integrative framework. *Journal of Environmental Psychology* 15(3): 169–182.

Karst, J., M. D. Jones, and J. D. Hoeksema. 2023. Positive citation bias and overinterpreted results lead to misinformation on common mycorrhizal networks in forests. *Nature Ecology & Evolution* 7(4): 501–511.

Kawai, M. 1965. Newly-acquired pre-cultural behavior of the natural troop of Japanese monkeys on Koshima Islet. *Primates* 6: 1–30.

Kellert, S. R., D. J. Case, D. Escher, D. J. Witter, J. Mikels-Carrasco, and P. T. Seng. 2017. *The Nature of Americans. National Report*. Available at https://natureofamericans.org/sites/default/files/reports/Nature-of-Americans_National_Report_1.3_4-26-17.pdf.

Kimball, J. W. 1944. A fishy bird story. *Auk* 61: 646–647.

Kinsey, A. C. 1920. Life histories of American Cynipidae. *Bulletin of the American Museum of Natural History* 42(6): 319–357.

Kinsey, A. C. 1922. Studies of some new and described Cynipidae (Hymenoptera). Indiana University Studies 9(53): 3–141.

Klump, B. C., R. E. Major, D. R. Farine, J. M. Martin, and L. M. Aplin. 2022. Is bin-opening in cockatoos leading to an innovation arms race with humans? *Current Biology* 32(17): R910–R911.

Klump, B. C., J. M. Martin, S. Wild, J. K. Hörsch, R. E. Major, and L. M. Aplin. 2021. Innovation and geographic spread of a complex foraging culture in an urban parrot. *Science* 373(6553): 456–460.

Korshunova, T. A., T. Bakken, V. V. Grøtan, K. B. Johnson, K. Lundin, and A. V. Martynov. 2021. A synoptic review of the family Dendronotidae (Mollusca: Nudibranchia): A multilevel organismal diversity approach. *Contributions to Zoology* 90(1): 93–153.

Krombein, K. V., P. D. Hurd, D. R. Smith, and B. D. Burks. 1979. *Catalog of Hymenoptera in America North of Mexico*. Washington, DC: Smithsonian Institution Press.

Kyba, C. C., Y. Ö. Altıntaş, C. E. Walker, and M. Newhouse. 2023. Citizen scientists report global rapid reductions in the visibility of stars from 2011 to 2022. *Science* 379(6629): 265–268.

Kyba, C. C., T. Kuester, A. Sánchez de Miguel, K. Baugh, A. Jechow, F. Hölker, J. Bennie, et al. 2017. Artificially lit surface of Earth at night increasing in radiance and extent. *Science Advances* 3(11): e1701528.

Kyba, C. C., S. B. Pritchard, A. R. Ekirch, A. Eldridge, A. Jechow, C. Preiser, D. Kunz, et al. 2020. Night matters—Why the interdisciplinary field of "Night Studies" is needed. *J* 3(1): 1–6.

Laighton, A. 1878. *Poems*. Boston: A. Williams and Company.

Laland, K. N., and V. M. Janik. 2006. The animal cultures debate. *Trends in Ecology and Evolution* 21(10): 542–547.

Lambert, M. R., K. I. Brans, S. Des Roches, C. M. Donihue, and S. E. Diamond. 2021. Adaptive evolution in cities: Progress and misconceptions. *Trends in Ecology and Evolution* 36(3): 239–257.

Lane, N. 2015. The unseen world: Reflections on Leeuwenhoek (1677) "Concerning little animals." *Philosophical Transactions of the Royal Society B* 370: 20140344.

Larew, H. G. 1987. Oak galls preserved by the eruption of Mount Vesuvius in AD 79, and their probable use. *Economic Botany* 41(1): 33–40.

Lee, R. B. 1978. Ecology of a contemporary San people. In *The Bushmen: San Hunters and Herders of Southern Africa*, edited by P. V. Tobias, 98–114. Cape Town: Human and Rousseau.

Letourneau, D. K., and L. A. Dyer. 1998. Experimental test in lowland tropical forest shows top-down effects through four trophic levels. *Ecology* 79(5): 1678–1687.

Lewis-Phillips, J., S. Brooks, C. D. Sayer, R. McCrea, G. M. Siriwardena, and J. C. Axmacher. 2019. Pond management enhances the local abundance and species richness of farmland bird communities. *Agriculture, Ecosystems & Environment* 273: 130–140.

Łopucki, R., I. Kitowski, M. Perlińska-Teresiak, and D. Klich. 2021. How is wildlife affected by the COVID-19 pandemic? Lockdown effect on the road mortality of hedgehogs. *Animals* 11(3): 868.

Lowry, H., A. Lill, and B. B. Wong. 2013. Behavioural responses of wildlife to urban environments. *Biological Reviews* 88(3): 537–549.

MacDonald, K., F. Scherjon, E. van Veen, K. Vaesen, and W. Roebroeks. 2021. Middle Pleistocene fire use: The first signal of widespread cultural diffusion in human evolution. *Proceedings of the National Academy of Sciences* 118(31): e2101108118.

Martin, D., M. Mecca, M. A. Meca, G. van Moorsel, and C. Romano. 2022. Citizen science and integrative taxonomy reveal a great diversity within Caribbean Chaetopteridae (Annelida), with the description of one new species. *Invertebrate Systematics* 36(7): 581–607.

Mayne, R. C. 1862. *Four Years in British Columbia and Vancouver Island*. London: John Murray.

McAtee, W. L. 1926. The relation of birds to woodlots in New York State. *Roosevelt Wildlife Bulletin* 4(1): 7–152.

Mispagel, M. E., and S. D. Rose. 1978. *Arthropods Associated with Various Age Stands of Douglas-Fir from Foliar, Ground, and Aerial Strata*. Coniferous Forest Biome Bulletin No. 13. University of Washington, Seattle, 55 pp.

Mo, M., and E. Mo. 2022. Using the iNaturalist application to identify reports of green iguanas (*Iguana iguana*) on the mainland United States of America outside of populations in Florida. *Reptiles & Amphibians* 29(1): 85–92.

Montgomerie, R. D., and C. A. Redsell. 1980. A nesting hummingbird feeding solely on arthropods. *Condor* 82: 463–464.

Moor, H., A. Bergamini, C. Vorburger, R. Holderegger, C. Bühler, S. Egger, and B. R. Schmidt. 2022. Bending the curve: Simple but massive conservation action leads to landscape-scale recovery of amphibians. *Proceedings of the National Academy of Sciences* 119(42): e2123070119.

Moore, A. M. T., G. C. Hillman, and A. J. Legge. 2000. *Village on the Euphrates: From Foraging to Farming at Abu Hureyra*. Oxford: Oxford University Press.

Moore, M. P., K. Hersch, C. Sricharoen, S. Lee, C. Reice, P. Rice, S. Kronick, et al. 2021. Sex-specific ornament evolution is a consistent feature of climatic adaptation across space and time in dragonflies. *Proceedings of the National Academy of Sciences* 118(28): e2101458118.

Narango, D. L., D. W. Tallamy, and P. P. Marra. 2017. Native plants improve breeding and foraging habitat for an insectivorous bird. *Biological Conservation* 213: 42–50.

Narango, D. L., D. W. Tallamy, and P. P. Marra. 2018. Nonnative plants reduce population growth of an insectivorous bird. *Proceedings of the National Academy of Sciences* 115(45): 11549–11554.

Nash, L. A. 1890. Some memories of Charles Darwin. *Overland Monthly* 16: 404–408.

Neitlich, P. N. 1993. Lichen abundance and biodiversity along a chronosequence from young managed stands to ancient forest. Master's thesis, University of Vermont.

Nguyen, P. Y., T. Astell-Burt, H. Rahimi-Ardabili, and X. Feng. 2023. Effect of nature prescriptions on cardiometabolic and mental health, and physical activity: A systematic review. *The Lancet Planetary Health* 7(4): e313–e328.

Nguyen, T., M. Saleh, M. Kyaw, G. Trujillo, M. Bejarano, K. Tapia, D. Waetjen, et al. 2020. Special report 4: Impact of COVID-19 mitigation on wildlife-vehicle conflict. Road Ecology Center, University of California, Davis. https://roadecology.ucdavis.edu/sites/g/files/dgvnsk8611/files/files/COVID_CHIPs_Impacts_wildlife.pdf.

Nicholls, T. H., and M. R. Fuller. 1987. Territorial aspects of barred owl home range and behavior in Minnesota. In *Biology and Conservation of Northern Forest Owls: Symposium Proceedings. February 3–7, 1987, Winnipeg, Manitoba*, edited by R. W. Nero, R. J. Clark, R. J. Knapton, and R. H. Harmre, 121–128. General Technical Report RM-142. Fort Collins, CO: US Department of Agriculture, Forest Service, Rocky Mountain Forest and Range Experiment Station.

Nielsen, U. N., D. H. Wall, and J. Six. 2015. Soil biodiversity and the environment. *Annual Review of Environment and Resources* 40: 63–90.

Nietzsche, F. 1910. *The Joyful Wisdom*. Translated by T. Common. London: T. N. Foulis.

Nietzsche, F. 1974. *The Gay Science*. Translated by W. Kaufmann. New York: Vintage Books.

Nihei, Y., and H. Higuchi. 2001. When and where did crows learn to use automobiles as nutcrackers? *Tohoku Psychologica Folia* 60: 93–97.

Nisbett, A. 1976. *Konrad Lorenz*. New York: Harcourt Brace Jovanovich.

Nowak, D. J., S. Maco, and M. Binkley. 2018. i-Tree: Global tools to assess tree benefits and risks to improve forest management. *Arboricultural Consultant* 51(4): 10–13.

O'Connell, J. F., K. Hawkes, and N. B. Jones. 1988. Hadza scavenging: Implications for Plio/Pleistocene hominid subsistence. *Current Anthropology* 29(2): 356–363.

Oh, R. M., M. L. Neo, N. Yap, S. Jain, R. Tan, C. A. Chen, and D. Huang. 2019. Citizen science meets integrated taxonomy to uncover the diversity and distribution of Corallimorpharia in Singapore. *Raffles Bulletin of Zoology* 67: 306–321.

Öhman, A., and S. Mineka. 2001. Fears, phobias, and preparedness: Toward an evolved module of fear and fear learning. *Psychological Review* 108(3): 483–522.

Orfinger, A. B., D. Helsel, and S. F. Breeding. 2018. Longevity of the barred owl (*Strix varia* Barton, 1799) from captivity. *Wilson Journal of Ornithology* 130(5): 1009–1010.

Ossola, A., U. M. Irlich, and J. Niemelä. 2018. Bringing urban biodiversity research into practice. In *Urban Biodiversity*, edited by A. Ossola and J. Niemelä, 1–17. London: Routledge.

Owens, A. C. S., and S. M. Lewis. 2018. The impact of artificial light at night on nocturnal insects: A review and synthesis. *Ecology and Evolution* 8(22): 11337–11358.

Paine, R. T. 1966. Food web complexity and species diversity. *The American Naturalist* 100(910): 65–75.

Pandit, R., and D. N. Laband. 2010. Energy savings from tree shade. *Ecological Economics* 69: 1324–1329.

Partners in Flight. 2022. Population estimates database, version 3.1. Accessed November 28, 2022. Available at http://pif.birdconservancy.org/PopEstimates.

Paulson, D. 2009. *Dragonflies and Damselflies of the West*. Princeton, NJ: Princeton University Press.

Phillips, I. D., R. D. Vinebrooke, and M. A. Turner. 2009. Experimental reintroduction of the crayfish species *Orconectes virilis* into formerly acidified Lake 302S (Experimental Lakes Area, Canada). *Canadian Journal of Fisheries and Aquatic Sciences* 66(11): 1892–1902.

Phillpotts, E. 1918. *A Shadow Passes*. London: Cecil, Palmer, and Hayward.

Plummer, K. E., K. Risely, M. P. Toms, and G. M. Siriwardena. 2019. The composition of British bird communities is associated with long-term garden bird feeding. *Nature Communications* 10: 2088.

Porto, R., R. F. de Almeida, O. Cruz-Neto, M. Tabarelli, B. F. Viana, C. A. Peres, and A. V. Lopes. 2020. Pollination ecosystem services: A comprehensive review of economic values, research funding and policy actions. *Food Security* 12(2): 1425–1442.

Preston, R. 2007. *The Wild Trees*. New York: Random House.

Proust, M. 2006. *Remembrance of Things Past, Vol. 2*. Translated by C. K. S. Moncrieff and S. Hudson. Hertfordshire, UK: Wordsworth Editions.

Quiller-Couch, A. T. 1920. *On the Art of Reading*. London: Cambridge University Press.

Ripple, W. J., and R. L. Beschta. 2012. Trophic cascades in Yellowstone: The first 15 years after wolf reintroduction. *Biological Conservation* 145(1): 205–213.

Ripple, W. J., R. L. Beschta, and L. E. Painter. 2015. Trophic cascades from wolves to alders in Yellowstone. *Forest Ecology and Management* 354: 254–260.

Robertson, L. 2019. Was Antoni van Leeuwenhoek secretive? His experiments with insect corneas. *FEMS Microbiology Letters* 366(16): fnz194.

Robles, A. J., L. Du, R. H. Cichewicz, and S. L. Mooberry. 2016. Maximiscin induces DNA damage, activates DNA damage response pathways, and has selective cytotoxic activity against a subtype of triple-negative breast cancer. *Journal of Natural Products* 79(7): 1822–1827.

Rodrigues, T. M., E. H. Miller, S. V. Drovetski, R. M. Zink, J. Fjeldså, and D. Gonçalves. 2021. Phenotypic divergence in two sibling species of shorebird: Common Snipe and Wilson's Snipe (Charadriiformes: Scolopacidae). *Ibis* 163(2): 429–447.

Roesch, L. F., R. R. Fulthorpe, A. Riva, G. Casella, A. K. M. Hadwin, A. D. Kent, S. H. Daroub, et al. 2007. Pyrosequencing enumerates and contrasts soil microbial diversity. *The ISME Journal* 1: 283–290.

Rosa, C. D., and S. Collado. 2019. Experiences in nature and environmental attitudes and behaviors: Setting the ground for future research. *Frontiers in Psychology* 10: 763.

Rule, S., B. W. Brook, S. G. Haberle, C. S. M. Turney, A. P. Kershaw, and C. N. Johnson. 2012. The aftermath of megafaunal extinction: Ecosystem transformation in Pleistocene Australia. *Science* 335(6075): 1483–1486.

Ruppert, K. M., R. J. Kline, and M. S. Rahman. 2019. Past, present, and future perspectives of environmental DNA (eDNA) metabarcoding: A systematic review in methods, monitoring, and applications of global eDNA. *Global Ecology and Conservation* 17: e00547.

Russo, G., A. Milks, D. Leder, T. Koddenberg, B. M. Starkovich, M. Duval, J. X. Zhao, et al. 2023. First direct evidence of lion hunting and the early use of a lion pelt by Neanderthals. *Scientific Reports* 13: 16405.

Sanders, D., E. Frago, R. Kehoe, C. Patterson, and K. J. Gaston. 2021. A meta-analysis of biological impacts of artificial light at night. *Nature Ecology & Evolution* 5: 74–81.

Schatz, A. 1993. The true story of the discovery of streptomycin. *Actinomycetes* 4(2): 27–39.

Schilthuizen, M., L. A. Seip, S. Otani, J. Suhaimi, and I. Njunjić. 2017. Three new minute leaf litter beetles discovered by citizen scientists in Maliau Basin, Malaysian Borneo (Coleoptera: Leiodidae, Chrysomelidae). *Biodiversity Data Journal* 5: e21947.

Schlots, F. E., A. O. Ness, J. J. Rasmussen, C. J. McMurphy, L. L. Main, R. J. Richards, W. A. Starr, et al. 1962. *Soil Survey of San Juan County, Washington.* Washington, DC: US Government Printing Office.

Schmidt, K. J., D. A. Soluk, S. E. M. Maestas, and H. B. Britten. 2021. Persistence and accumulation of environmental DNA from an endangered dragonfly. *Scientific Reports* 11: 18987.

Schmitz, O. J., and S. J. Leroux. 2020. Food webs and ecosystems: Linking species interactions to the carbon cycle. *Annual Review of Ecology, Evolution, and Systematics* 51: 271–295.

Schmitz, O. J., M. Sylvén, T. B. Atwood, E. S. Bakker, F. Berzaghi, J. F. Brodie, J. P. G. M. Cromsigt, et al. 2023. Trophic rewilding can expand natural climate solutions. *Nature Climate Change* 13: 324–333.

Schultes, R. E., and R. F. Raffauf. 1990. *The Healing Forest: Medicinal and Toxic Plants of the Northwest Amazonia.* Portland, OR: Dioscorides Press.

Shin, Y., K. Kim, J. Groffen, D. Woo, E. Song, and A. Borzée. 2022. Citizen science and roadkill trends in the Korean herpetofauna: The importance of spatially biased and unstandardized data. *Frontiers in Ecology and Evolution* 10: 944318. https://doi.org/10.3389/fevo.2022.944318.

Simard, S. W., K. J. Beiler, M. A. Bingham, J. R. Deslippe, L. J. Philip, and F. P. Teste. 2012. Mycorrhizal networks: Mechanisms, ecology and modelling. *Fungal Biology Reviews* 26: 39–60.

Solomon, R. 2022. With so many undiscovered mushrooms, citizen scientists find new species all the time. National Public Radio. Archived at www.npr.org/2022/09/22/1124590354/. Accessed October 13, 2022.

Starnberger, I., D. Preininger, and W. Hödl. 2014. The anuran vocal sac: A tool for multimodal signalling. *Animal Behaviour* 97: 281–288.

Starr, G. D., T. J. Davis, and J. A. V. Quintanilla. 2021. A cryptic new species of *Agave* (Asparagaceae/Agavoideae) and an amplified description of *Agave tenuifolia*. *Cactus and Succulent Journal* 93(4): 273–285.

Steinke, D., V. Breton, E. Berzitis, and P. D. Hebert. 2017. The School Malaise Trap Program: Coupling educational outreach with scientific discovery. *PLoS Biology* 15(4): e2001829.

Stevenson, M. P., T. Schilhab, and P. Bentsen. 2018. Attention Restoration Theory II: A systematic review to clarify attention processes affected by exposure to natural environments. *Journal of Toxicology and Environmental Health, Part B* 21(4): 227–268.

Swift, J. 2010. *Gulliver's Travels*. London: Penguin Classics.

Tanaka, K. R., K. S. Van Houtan, E. Mailander, B. S. Dias, C. Galginaitis, J. O'Sullivan, C. G. Lowe, et al. 2021. North Pacific warming shifts the juvenile range of a marine apex predator. *Scientific Reports* 11: 3373.

Teale, E. W. 1974. *A Naturalist Buys an Old Farm*. New York: Ballantine Books.

Terashima, H., and M. Ichikawa. 2003. A comparative ethnobotany of the Mbuti and Efe hunter-gatherers in the Ituri forest,

Democratic Republic of Congo. *African Study Monographs* 24(1, 2): 1–168.

Thoreau, H. D. 1966. *Walden and Civil Disobedience*. New York: W. W. Norton and Company.

Tilman, D., P. B. Reich, and J. M. Knops. 2006. Biodiversity and ecosystem stability in a decade-long grassland experiment. *Nature* 441: 629–632.

Tolkien, J. R. R. 2012. *The Hobbit*. Boston: Houghton Mifflin Harcourt.

Treat, M. 1885. *Home Studies in Nature*. New York: Harper and Brothers.

Treves, A., and P. Palmqvist. 2007. Reconstructing hominin interactions with mammalian carnivores (6.0–1.8 Ma). In *Primate Anti-predator Strategies*, edited by S. Gurskey-Doyen and K. A. I. Nekaris, 355–381. Boston, MA: Springer.

Turner, C. H. 1908. The homing of the burrowing-bees (Anthophoridae). *Biological Bulletin* 15(6): 247–258.

Turner, C. H. 1909. The behavior of a snake. *Science* 30: 563–564.

Twining, C. W., J. R. Shipley, and D. W. Winkler. 2018. Aquatic insects rich in omega-3 fatty acids drive breeding success in a widespread bird. *Ecology Letters* 21(12): 1812–1820.

Tyler, W. M. 1914. Notes on nest life of the brown creeper in Massachusetts. *Auk* 31(1): 50–62.

Ueda, K. 2020. iNaturalist as a tool for citizen science. Sonoma Land Trust webinar. Archived at www.youtube.com/watch?v=520CMdxuyDc&t=1866s. Accessed October 19, 2022.

Ulrich, R. S. 1984. View through a window may influence recovery from surgery. *Science* 224(4647): 420–421.

Usherwood, J. R., E. L. Sparkes, and R. Weller. 2014. Leap and strike kinetics of an acoustically "hunting" barn owl (*Tyto alba*). *Journal of Experimental Biology* 217(17): 3002–3005.

van Achterberg, K., M. Schilthuizen, M. van der Meer, R. Delval, C. Dias, M. Hoynck, H. Köster, et al. 2020. A new parasitoid wasp, *Aphaereta vondelparkensis* sp. n. (Braconidae, Alysiinae) from a city park in the centre of Amsterdam. *Biodiversity Data Journal* 8: e49017.

van der Ploeg, R. R., W. Böhm, and M. B. Kirkham. 1999. On the origin of the theory of mineral nutrition of plants and the law of the minimum. *Soil Science Society of America Journal* 63(5): 1055–1062.

van Leeuwenhoek, A. 1702. Part of a letter from Mr Antony van Leeuwenhoek, F. R. S. concerning green weeds growing in water, and some *Animalcula* found about them. *Philosophical Transactions of the Royal Society of London* 23(283): 1304–1311.

van Leeuwenhoek, A. 1979. *Alle de Brieven. Deel 10: 1694–1695.* Edited by L. C. Palm. Amsterdam, Netherlands: N.V. Swets and Zeitlinger.

Wadman, M. 2020. The long shot. *Science* 370(6517): 649–653.

Wagg, C., S. F. Bender, F. Widmer, and M. G. van der Heijden. 2014. Soil biodiversity and soil community composition determine ecosystem multifunctionality. *Proceedings of the National Academy of Sciences* 111(14): 5266–5270.

Wallace, A. R. 1872. *The Malay Archipelago*. London: Macmillan and Company.

Wangberg, J. K. 1975. Biology of the thimbleberry gallmaker *Diastrophus kincaidii* (Hymenoptera: Cynipidae). *The Pan-Pacific Entomologist* 51(1): 39–48.

Wangberg, J. K. 1976. The insect community in galls of *Diastrophus kincaidii* Gillette (Hymenoptera: Cynipidae) on thimbleberry. University of Idaho Department of Entomology 50th Anniversary Publication 7: 45–50.

Wege, M. L., and D. G. Raveling. 1983. Factors influencing the timing, distance, and path of migrations of Canada geese. *Wilson Bulletin* 95(2): 209–221.

Whiten, A. 2021. The burgeoning reach of animal culture. *Science* 372(6537): eabe6514.

Whitman, W. 1976. *Leaves of Grass*. Secaucus, NJ: Longriver Press.

Williams, C. K., M. D. Samuel, V. V. Baranyuk, E. G. Cooch, and D. Kraege. 2008. Winter fidelity and apparent survival of lesser snow goose populations in the Pacific flyway. *Journal of Wildlife Management* 72(1): 159–167.

Wilson, E. O. 1987. The little things that run the world (the importance and conservation of invertebrates). *Conservation Biology* 1(4): 344–346.

Yamazaki, K. 2016. Caterpillar mimicry by plant galls as a visual defense against herbivores. *Journal of Theoretical Biology* 404: 10–14.

Zhang, C. L. Wen, Y. Wang, C. Liu, Y. Zhou, and G. Lei. 2020. Can constructed wetlands be wildlife refuges? A review of their potential biodiversity conservation value. *Sustainability* 12(4): 1442.

Zhang, H., J. Yang, L. Zhang, X. Gu, and X. Zhang. 2023. Citizen science meets eDNA: A new boom in research exploring urban wetland biodiversity. *Environmental Science and Ecotechnology* 16: 100275.

Zimov, S. A., N. S. Zimov, A. N. Tikhonov, and F. S. Chapin III. 2012. Mammoth steppe: A high-productivity phenomenon. *Quaternary Science Reviews* 57: 26–45.

Index

Aargau, Switzerland, 204–208
action, 9. *See also* community data collection
adaptive behaviors, 71–88
  bill pouncing, 71–72
  of burrowing-bees, 76–77
  challenges in studying, 81
  of cliff swallows, 88
  of cockatoos, 78–84, 80 (fig.), 87
  and counteradaptation, 83
  of crows, 83–84
  and culture development, 82
  and evolution, 88
  foraging, 74, 78–84
  homing, 76–77
  and human learning, 83
  of mice, 88
  of murmurations, 85–86, 87 (fig.)
  notes on robin and frog, 71–74, 73 (fig.)
  nut dropping, 84
  of prey droppers, 74
  #StarlingSurvey observations, 86
  of starlings, 85–86, 87 (fig.)
  and sustained observation, 75
  tool use in, 84
  in urban settings, 87–88
  See also backyard wildlife observation
Agassiz, Louis, 19–20
À Giverny, chez Claude Monet (Monet), 129
amateur naturalists. *See* community data collection
amphibians, 204–208
Anthony, Mark, 116–118
the anthropause, 4–5
antibiotics, soil derived, 127
Antibiotics Unearthed project, 128
aphids, 97–102, 99 (fig.)
Apophthegms, Sentiments, Opinions and Occasional Reflections (Johnson), 149
archerfish, 139
artificial light, 161–166, 164 (fig.)
attentiveness, 31. *See also* observation
auditory awareness, 157–160
augury, 171

backyard wildlife observation
and adaptive behaviors, 71–88
and biodiversity conservation, 211–222
of fungi, 120–128
and habitat recovery, 191–209
of nocturnal wildlife, 149–166
of pond life, 129–147
process, 13–47
and rewilding, 169–190
in soil, 109–128
species discovery in, 51–70
of trees, 89–107
bark beetles, 118
bark installations, 194–197, 197 (fig.), 202–203
barred owls, 149–153, 151 (fig.)
Bashō, Matsuo, 211
bees
black mud bees, 32–33, 33 (fig.)
burrowing, 76–77
conservation of, 212–214
in Great Sunflower Project, 42–46, 44 (fig.)
navigation of, 76–77
notes on bee garden, 221–222
restoration of, 212–214, 221–222
tracking, 6–8
beetles, snout, 37
Beggs, Jacqueline, 102
Bent, Arthur Cleveland, 152
Beumer, Carijn, 213
Bibarrambla allenella (moth), 184
bill pouncing, 71–72
BIMBY (biodiversity in my back yard), 213
biodiversity, backyard
and animal migration, 62–63
countering loss of, 9
and the democratization of science, 65
eDNA tool for observing, 145–146
increasing, 164–166, 170
indicators of, 41
in soil, 112, 116–118
in urban settings, 88
value of, 105
See also backyard wildlife observation; community data collection; habitat recovery; rewilding, backyard; species discovery
biodiversity conservation, reasons for, 211–222
for bees, 212–214
for butterflies, 212
limits to human understanding, 217–220
monetary value, 216–217
notes on bee garden, 221–222
stewardship, 215
wellbeing for humans, 220–221
for whales, 218–219

for wolves, 218
See also backyard wildlife observation; habitat recovery; species discovery
biodiversity in my back yard (BIMBY), 213
biomimicry, 217
bioprospecting, 216–217
BioSCAN project, 52–56, 53 (fig.)
bird boxes, 199–202, 201 (fig.)
black mud bees, 32–33, 33 (fig.)
Blyth, Julia, 176–184
and galls, 181–182
and leaf miners, 179–180
and moth preservation, 183
strategies for backyard habitat restoration, 177–178
Book of Documents (Gong), 111
Brewer, Thomas Mayo, 193–194
Brown, Brian, 51–54, 69
brown creepers, 191–199, 194 (fig.)
diet of, 192
early appearance of, 191–192
feeding young, 198
limiting factor for, 193
nesting sites of, 194–199, 197 (fig.)
See also woodpeckers
brush piles, 135. *See also* observation
Buckler, Joe, 199–202
BugGuide.net, 178
Burroughs, John, 3, 13
burrowing-bees, 76–77
burrowing bugs, 118
butterflies, 184–189, 186 (fig.), 212
Bwindi Impenetrable National Park, 6–8

cameras, digital, 5, 34, 35, 62–65, 67. *See also* smartphones
cancer, 123–124, 126, 163, 221. *See also* medicinal species
canopies, forest. *See* trees
caterpillars, 95–96, 186
Cervantes, Miguel de, 71
Cheerios for fungi, 123
chemical extract production, 124, 126
children, 30–32, 37–38
Cichewicz, Robert, 122–128
citizen science projects. *See* community data collection
cliff swallows, 88
climate change, 62, 66–67, 188
cockatoos, 78–84, 80 (fig.), 87
colonizing behaviors, 179–180, 205, 208
community data collection
for adaptive behaviors, 88
for bee activity, 43–46
for cockatoo observations, 80–81
as educational for families, 56
and monitoring, 206
for nocturnal insects, 42–43
for pond life, 146

community data collection (*continued*)
for soil fungi, 122, 124–126, 128
for starling murmurations, 85–86, 87 (fig.)
See also backyard wildlife observation; biodiversity, backyard; observation; rewilding, backyard; species discovery; tools
complete habitats, 212–214
counteradaptation, 83
COVID-19 lockdowns, 4–5
crawling for observation, 38–40
creepers, brown, 191–199, 194 (fig.)
diet of, 192
early appearance of, 191–192
feeding young, 198
limiting factor for, 193
nesting sites of, 194–199, 197 (fig.)
See also woodpeckers
crepuscular species, 153. *See also* nocturnal wildlife
crows, 83–84
culture development, 82

daisy fleabane, 179–180, 180 (fig.)
darkness, human fear of, 153–155
Darwin, Charles, 21–27, 111
Darwin, Francis, 32
digital cameras, 5, 34, 35, 62–65, 67. *See also* smartphones
Diptera order. *See* flies
dispersal-limited species, 117
diurnal species, 153. *See also* nocturnal wildlife
Dokuchaev, Vasily, 111
Don Quixote (Cervantes), 71
Douglas firs, 24–25, 92, 93 (fig.), 195–196
Down House, 21–24
dragonflies, 140–142, 141 (fig.), 144–145, 146
Drugs from Dirt program, 128
Du Bois, W. E. B., 78
duckweed, 143 (fig.)

eBird, 65
ecological complexity, 217–220
ecotours, taxonomy-themed, 61
eDNA (environmental DNA), 145–146
Effe and Mbuti peoples, 5–6
Eiseman, Charley, 176–184
and *Bibarrambla allenella*, 183–184
and daisy fleabane, 179
and galls, 181–182
and leaf miners, 179–180
and sawflies, 182–183
strategies for backyard habitat restoration, 177–178
Eliot, George, 191
eluviation horizons, 114
empathy for species, 117–118

Engstrom, Brett, 110
environmental DNA (eDNA), 145–146
Epistles (Horace), 169
Erwin, Terry, 90–91
Ethier, Jeffrey, 157–158
Eumenins, 29–30, 34–37, 36 (fig.)
evolution, 67, 88
extinction, 219

Fabre, Jean-Henri, 32–34, 37
farmland habitats compared to suburban, 189
Fern, Fanny, 29
Fertilization of Orchids (C. Darwin), 22
fission-fusion system of foraging, 79
flashlights, 156–157, 160
fleabane, 179–180, 180 (fig.)
flies, 55–60, 60 (fig.), 68–69, 189
foraging behaviors, 78–84
foraging theory, 74
forest bathing, 221. *See also* nature therapy; trees
forests. *See* trees
foxes, 38–39
frogs, 71–74, 73 (fig.), 156–160, 159 (fig.)
fungi, 120–128
  Antibiotics Unearthed project, 128
  changing understanding of, 121
  Cheerios for, 123
  chemical extract production of, 124, 126
  Drugs from Dirt program, 128
  maximiscin derived from, 126
  for medicinal use, 123–124, 126–127 (*see also* cancer)
  pericosine derived from, 126–127
  pervasiveness of, 120–121
  sample of, 125 (fig.)
  streptomycin derived from, 127
  See also backyard wildlife observation; soil; species discovery

galls, 170–176, 172 (fig.), 181
gardens, 188
GBIF (Global Biodiversity Information Facility), 65
Geest, Emily, 184–188
Ginger-Snaps (Fern), 29
Global Biodiversity Information Facility (GBIF), 65
GLOBE Observer app, 104
Gong, Z., 111
Gonzalez, Lisa, 56
Goodenough, Anne, 85–86
Grahame, Kenneth, 109
Great Sunflower Project, 42–46, 44 (fig.)
green spaces, 220–221. *See also* nature therapy

growth control for habitat recovery, 193
Gulliver's Travels (Swift), 38

habitat recovery, 191–209
  affecting multiple species, 208
  for amphibians, 204–208
  bark installation for, 194–197, 197 (fig.), 202–203
  bird boxes for, 199–202, 201 (fig.)
  for brown creepers, 191–199, 194 (fig.), 197 (fig.)
  and growth control, 193
  indicators of need for, 193, 195, 198, 202, 203
  and law of the minimum, 193
  and limiting factors, 193, 195, 198, 202, 203
  metapopulations' effect on, 205
  monitoring, 206
  nests for, 191–203
  notes on brown creepers, 198
  notes on nests, 202–203
  pond construction for, 204–208, 207 (fig.)
  See also backyard wildlife observation; biodiversity, backyard; biodiversity conservation, reasons for; rewilding, backyard; species discovery
habituation, 137
Hanson, Thor, 6 (fig.), 93 (fig.)
Hartop, Emily, 54–58, 68, 69n
Henry IV, Part II (Shakespeare), 1
herbivory and predation patterns, 219
hermit thrushes, 1–3, 3 (fig.)
Hesse, Hermann, 102
Hine's dragonflies, 146
The Hobbit (Tolkien), 51
Homegrown National Park, 213
Homes for Sale (business), 199–202, 201 (fig.)
Home Studies in Nature (Treat), 49
homing behaviors, 76–77
honeydew, 99–102
Hooke, Robert, 145
Horace, 169
horizons, soil, 113–114, 114 (fig.)
horse arenas, 212–213
hospital bed study, 220–221. *See also* nature therapy
human power, 215
human understanding, limits to, 217–220
human wellbeing, 220–221
Hutton, James, 111
Hutton's vireo, 13–18, 15 (fig.)
hyperparasites, 176

iNaturalist website, 62–68
insects. *See specific insects*
i-Tree tool, 104–107

Johnson, Samuel, 149
The Joyful Wisdom (Nietzsche), 89

Kinsey, Alfred, 171–173, 175–176
Klump, Barbara, 78–84, 87, 88

law of the minimum, 193
Leaf and Tendril (Burroughs), 13
leaf miners, 178–181, 180 (fig.)
Leafminers of North America (Eiseman), 178
learning, human, 66, 83
LeBuhn, Gretchen, 42–46
Leeuwenhoek, Antonie van, 143–145
lichen, 14, 91–92
Liebig, Justus von, 193
light pollution, 161–166, 164 (fig.)
lightsheeting, 41–42
limiting factors, 193, 195, 198, 202, 203
limits to human understanding, 217–220
Loarie, Scott, 62–66, 68, 69
Lorenz, Konrad, 75

macrophotography, 119
Malaise, René, 57
Malaise traps, 57, 58
manna from heaven, 101
mason wasps, 29–30, 34–37, 36 (fig.)
Matthews, Phil, 140, 142
maximiscin, 126
Mbuti and Effe peoples, 5–6
medicinal species, 123–124, 126, 127, 171, 216–217. *See also* cancer
melatonin, 163
metapopulations, 205
mice, 88
Mickey Mouse rationale, 217–220. *See also* biodiversity conservation, reasons for
microorganisms, pond, 142–144, 143 (fig.)
Middlemarch (Eliot), 191
milkweed, 185–188, 186 (fig.)
monarch butterflies, 184–189, 186 (fig.)
Monet, Claude, 129
monetary value of nature, 216–217
monitoring habitat recovery, 206. *See also* community data collection
moths, 40–42, 164 (fig.), 181, 183, 184
murmurations, starling, 85–86, 87 (fig.)

naming, 69
National Cancer Institute (NCI), 123–124
Natural History (Pliny the Elder), 94
natural history road trips, 135

nature, people's connection to, 2–4
nature-based industries, 216
nature therapy, 31, 220–221
navigation behaviors, 76–77, 155
NCI (National Cancer Institute), 123–124
nesting habitats
  bee navigation to, 76–77
  bird boxes for, 199–202, 201 (fig.)
  of black mud bees, 33, 33 (fig.)
  of brown creepers, 192–199, 197 (fig.)
  cultivating, 194–203
  ground, 212–213, 222
  locating, 7–8, 33, 160
  of potter wasps, 35–37, 36 (fig.)
  of racoons, 17 (fig.)
  of vireos, 14–18, 15 (fig.)
  See also backyard wildlife observation; biodiversity, backyard; rewilding, backyard; species discovery
Nietzsche, Friedrich, 89
nocturnal wildlife, 149–166
  auditory awareness of, 157–160
  barred owls, 149–153, 151 (fig.)
  crepuscular species compared with, 153
  and darkness, human fear of, 153–155
  flashlights for observing, 156–157, 160
  frogs, 156–160, 159 (fig.)
  light pollution's effect on, 161–166, 164 (fig.)
  and melatonin, 163
  navigation behaviors of, 155
  snow geese, 155
  territorial behaviors of, 152
  "vacuum cleaner effect" on, 166
  See also backyard wildlife observation; biodiversity, backyard; community data collection; observation; species discovery
Nowak, David, 104–106
nut dropping, 84

observation, 13–47
  acoustic awareness, 157–160
  and attentiveness, 31
  brush piles for, 135
  by children, 30–33, 37–38
  Darwin's habits of, 21–23, 26–27
  digital cameras in, 5, 34, 35, 62–65, 67
  and habituation, 137
  by hunter-gatherers, 5–8
  as learned ability, 31
  lightsheeting use in, 41–42
  notes from author's backyard, 13–18, 24–28
  perspective's effect on, 38–40
  repetition in, 19–21

scaling of apparent distance for, 38
senses involved in, 38–39
sitting still for, 136
slowing down for, 18–21, 26, 75
volunteer participation in (*see* community data collection)
See also backyard wildlife observation; biodiversity, backyard; species discovery; tools
Odynerus genus, 29–30, 34–37, 36 (fig.)
On the Art of Reading (Quiller-Couch), 167
outdoor activity, human, 2–4
ovipositors, 35
owls, 149–153, 151 (fig.)

Pacific Northwest climate, 24
parasites, 175–176
parent material in soil, 114–115
parrots (cockatoos), 78–84, 80 (fig.), 87
pathfinders (wildflower), 39
path tortuosity, 161
patterns in nature, 67, 219
perception, filtering of, 129–130
pericosine, 126–127
Perkowski, Tawm, 90, 93–95
perspective, 38–40, 94, 103
pesticides, 45
Phillpotts, Eden, 11
phorids. *See* flies
photos, data uses of. *See* digital cameras
Pliny the Elder, 94
plume moths, 181
pollinators, 45. *See also specific pollinators*
pond construction, 204–208, 207 (fig.)
pond life, 129–147
archerfish, 139
dragonflies, 140–142, 141 (fig.), 144–145, 146
duckweed, 143 (fig.)
eDNA, 145–146
microorganisms, 142–144, 143 (fig.)
notes about, 130–134, 138–139, 147
rails, 131–134
snipes, 130–134, 131 (fig.)
See also backyard wildlife observation; biodiversity, backyard; community data collection; habitat recovery
population sinks, 187
Potter, Beatrix, 171
potter wasps, 29–30, 34–37, 36 (fig.)
power, human, 215
predation and herbivory patterns, 219
preservation techniques for moths, 183

prey droppers, 74
property values, 217
Proust, Marcel, 28

Quiller-Couch, Sir Arthur Thomas, 167

raccoons, 16–18, 17 (fig.)
Raccoon Shack (author's office), 27
rails, 131–134
refuges, backyard habitats as, 188
Rennels, Candice, 184, 185, 189
repetition in observation, 19–21
restoration. *See* habitat recovery; rewilding, backyard
rewilding, backyard, 169–190
  BugGuide.net for, 178
  daisy fleabane, 179–180, 180 (fig.)
  galls, 170–176, 172 (fig.), 181
  gardens, 188
  leaf miners, 178–181, 180 (fig.)
  milkweed, 185–188, 186 (fig.)
  monarch butterflies, 184–189, 186 (fig.)
  moths, 181, 183, 184
  and population sinks, 187
  for refuges, 188
  sawflies, 182
  suburban compared to farmlands, 189
  tachinid flies, 189
  thimbleberry canes, 169–170, 172 (fig.)
  and umbrella species, 188–189
  wasps, 172–176, 172 (fig.)
  See also backyard wildlife observation; biodiversity, backyard; biodiversity conservation, reasons for; community data collection; habitat recovery; species discovery
road trips, natural history, 135
robins, 71–74, 73 (fig.)
Royal Horticultural Society, 24

Sand-walk, 22, 23 (fig.)
San hunters, 5
sawflies, 182
scaling of apparent distance, 38
Schatz, Albert, 127
Schmidt, Benedikt, 204–209
Schroer, Sibylle, 163, 165
scorpions, 67
Scrooge McDuck rationale, 216–217. *See also* biodiversity conservation, reasons for
Scudder, Samuel H., 19–20
senses for observation, 38–39

Sexual Behavior in the Human Male (Kinsey), 171
A *Shadow Passes* (Phillpotts), 11
Shakespeare, William, 1
sitting still for observation, 136
skewed results, 56
slowing down for observation, 18–21, 26
smartphones, 64, 65–66. *See also* digital cameras
snipes, 130–134, 131 (fig.)
snout beetles, 37
snow geese, 155
soil, 109–128
  biodiversity in, 112, 116–118
  burrowing bugs in, 118
  classification systems for, 111
  dispersal-limited species in, 117
  eluviation horizons in, 114
  empathy for species of, 117–118
  formation of, 115
  fungi in, 120–128, 125 (fig.)
  horizons of, 113–114, 114 (fig.)
  macrophotography for, 119
  organisms in, 119 (fig.) (*see also* fungi)
  parent material in, 114–115
  varieties of, 110–111
  See also backyard wildlife observation; biodiversity, backyard; community data collection; species discovery
species, value of, 215
species discovery, 51–70
  automatic identification of, 64, 65–66
  biodiversity in, 62–63, 69
  by BioSCAN project, 52–56, 53 (fig.)
  and climate change, 62, 66–67
  digital camera use for, 34, 35, 62, 63, 64–65, 67
  by eBird, 65
  and evolution, 67
  flies, 52–56, 58–60, 60 (fig.), 68–69
  and human learning, 66
  by iNaturalist website, 62–68
  Malaise traps for, 57, 58
  notes on insects, 58–60
  pattern detection for, 67
  sampling tools for, 57, 58 (*see also* tools)
  scorpions, 67
  and skewed results, 56
  smartphone use for, 64, 65–66
  taxonomic impediment in, 61, 68–69
  See also backyard wildlife observation; biodiversity, backyard; biodiversity conservation, reasons for; community data collection; habitat recovery; rewilding, backyard

species identification,
automatic, 64, 65–66
Spider-Man rationale, 215–216.
See also biodiversity
conservation, reasons for
spiderwebs, 40
Sprengel, Carl, 193
#StarlingSurvey, 86
starlings, 85–86, 87 (fig.)
stewardship, 215
streptomycin, 127
suburban habitats compared to
farmland, 189
sunflowers, 42–46, 44 (fig.)
sunlight, 211
Swift, Jonathan, 38

tachinid flies, 189. *See also* flies
The Tale of Squirrel Nutkin
(Potter), 171
Tallamy, Douglas, 40–42,
46–47, 213
Taxon Expeditions, 61
taxonomic impediment, 61,
68–69. *See also* community
data collection
Teale, Edwin Way, 134–136,
138
technology. *See* tools
territorial behaviors, 152
Theory of the Earth (Hutton), 111
thimbleberry canes, 169–170,
172 (fig.)
Thoreau, Henry David, 8
Thorley, Catherine, 26–27
thrushes, 1–3, 3 (fig.)
Tolkien, J. R. R., 51
tools
for adaptive behaviors, 84
digital cameras, 5, 34, 35,
62–65, 67
eBird, 65
eDNA, 145–146
flashlights, 156–157, 160
GLOBE Observer app, 104
iNaturalist website, 62–68
i-Tree, 104–107
lightsheeting, 42, 43 (fig.)
Malaise traps, 57, 58
smartphones, 64, 65–66
#StarlingSurvey, 86
TreeSnap program, 104
Treezilla project, 103
See also backyard wildlife
observation; community
data collection;
observation
traps for flies, 57, 58. *See also*
flies
Treat, Mary, 49
trees, 89–107
accessing, 90–91
aphids in, 97–102, 99 (fig.)
caterpillars in, 95–96
Douglas firs, 24–25, 92, 93
(fig.), 195–196
forest bathing, 221
GLOBE Observer app, 104
i-Tree tool, 104–107
lichen in, 14, 91–92
monetary value of, 104
notes about, 24–26

perspective from, 94, 103
TreeSnap program, 104
Treezilla project, 103
wasps in, 94–98, 99 (fig.), 100–102
See also backyard wildlife observation; biodiversity, backyard; observation; species discovery
TreeSnap program, 104
Treezilla project, 103
Turner, Charles Henry, 76–78
turret builders (wasp), 29–30, 34–37, 36 (fig.)
Tyler, Winsor, 197

Ueda, Ken-ichi, 63
umbrella species, defined, 188–189
urban ecology, 87–88

vacuum cleaner effect, 166
vireos, 13–18, 15 (fig.)
Virginia rails, 131–134
"Vitality of Seeds" (C. Darwin), 22
vitamin D, 211
volunteers. *See* community data collection

Waksman, Selman, 127
Walden Pond, 8
Wallace, Alfred Russel, 31
wasps
as biomimicry sources, 217
gall wasps, 172–176, 172 (fig.)
potter wasps, 29–30, 34–37, 36 (fig.)
yellowjackets, 94–103, 99 (fig.)
Webb, Emma, 184, 185, 188–189
wellbeing, human, 220–221
whales, 218–219
"When Lilacs Last in the Dooryard Bloom'd" (Whitman), 2
Whitman, Walt, 2
Wilson, E. O., 59
Wilson's snipes, 130–134, 131 (fig.)
The Wind in the Willows (Grahame), 109
wolves, 218
woodpeckers, 191–199, 194 (fig.)
diet of, 192
early appearance of, 191–192
feeding young, 198
limiting factor for, 193
nesting sites of, 194–199, 197 (fig.), 200

yellowjackets, 94–103, 99 (fig.)
and aphids, 97–102
and caterpillars, 95–96
eyes of, 96–97
and honeydew, 99–102
See also wasps
Yu Gong (Gong), 111

Credit: Chase Anderson

Thor Hanson is a conservation biologist, Guggenheim fellow, and author of award-winning books including *Hurricane Lizards and Plastic Squid*, *Buzz*, *Feathers*, and *The Triumph of Seeds*. He lives with his wife and son on an island in Washington State.